Thomas Straub

Experimental Investigation of Crack Initiation

in Face-Centered Cubic Materials in the High
and Very High Cycle Fatigue Regime

Schriftenreihe
des Instituts für Angewandte Materialien
Band 55

Karlsruher Institut für Technologie (KIT)
Institut für Angewandte Materialien (IAM)

Eine Übersicht aller bisher in dieser Schriftenreihe erschienenen Bände finden Sie am Ende des Buches.

Experimental Investigation of Crack Initiation

in Face-Centered Cubic Materials in the High and Very High Cycle Fatigue Regime

by
Thomas Straub

KIT Scientific Publishing

Dissertation, Karlsruher Institut für Technologie (KIT)
Fakultät für Maschinenbau
Tag der mündlichen Prüfung: 9. Dezember 2015

Impressum

KIT Scientific Publishing

Karlsruher Institut für Technologie (KIT)
KIT Scientific Publishing
Straße am Forum 2
D-76131 Karlsruhe

KIT Scientific Publishing is a registered trademark of Karlsruhe Institute of Technology. Reprint using the book cover is not allowed.

www.ksp.kit.edu

Print on Demand 2016

ISSN 2192-9963
ISBN 978-3-7315-0471-9
DOI 10.5445/KSP/1000051515

Experimental Investigation of Crack Initiation in Face-Centered Cubic Materials in the High and Very High Cycle Fatigue Regime

Zur Erlangung des akademischen Grades

Doktor der Ingenieurwissenschaften

der Fakultät für Maschinenbau
Karlsruher Institut für Technologie (KIT)

genehmigte

Dissertation

von

M.Sc. Thomas Straub

Tag der mündlichen Prüfung:	9. Dezember 2015
Hauptreferent:	Prof. Dr. rer. nat. Oliver Kraft
1. Korreferent:	Prof. Dr. rer. nat. Christoph Eberl
2. Korreferent:	Prof. Dr. Yuri Lapusta

Table of contents

List of figures

List of tables

Acknowledgement

Most of the work presented in this thesis has been conducted from July 2009 to July 2013 at the Institute for Applied Materials at the Karlsruhe Institute of Technology, Germany. This work was then finished and written from August 2013 to November 2015 in my free time, at the same time as my full time job at the Fraunhofer Institute for Mechanics of Materials (IWM) in Freiburg, Germany. With this coming to an end, I would like to thank every person who contributed directly or indirectly to the success of this work.

Firstly, I would like to thank Prof. Oliver Kraft for being the primary examiner of this thesis, for his suggestions and contributions to this work, as well as for the continuous support of the group of Prof. Eberl.

My most sincere thanks go to Prof. Chris Eberl for his dedicated support of this work. Chris, you are the person without whom anything would have been possible. Who would have believed the trust you put in a French student in February 2008 would have ended in such a successful collaboration. Thank you for all your ideas, and the always positive and constructive discussions. For all that: high five! And for his role in the group, my thoughts also go to Dr. Tobias Kennerknecht. All those hours there and back again…

I would like to thank Prof. Yuri Lapusta from the IFMA (French Institute for Advanced Mechanics) in Clermont-Ferrand, France. Yuri, you gave me a long time ago the possibility to go to Germany and to start this unexpected journey. Merci. This also allowed me to go to the USA for the first time, where Assoc. Prof. Olivier Pierron gave me the chance to spend a short semester at the Georgia Institute of Technology, Atlanta, GA, USA. Best time ever, Olivier merci!

My thanks also go to Mr. Geoffroy Kieffer for his work on the developpment and the CAD of the first setup prototype, Mr. Michael Meidlinger for his work with Labview in his 'Diplomarbeit' as well as Mr. Ewald Ernst for his work at the workshop; Mrs. Daniela Exner for the briefings on the Dual Beam microscope; and Mr. Jochen Bach, Mr. Johannes Möller and Mr. Philipp Hilgendorff for our constructive discussions during the different SPP1466 meetings.

Some special thanks go to Mr. Mattweh Berwind for our work in 2012 on the setup and especially on the first observation of the frequency increase. I also would like to acknowledge Mr. Michael Buck for taking on the responsibility of continuing this project in the SPP1466.

A huge thank you goes to all former and current PhD students and colleagues for a pleasant working atmosphere as well as many funny times. Felix, Mélanie, Sofie, Matthias, Sven, Stefan, Nicola, Jochen, Manu, Irina, Julia, André, David, Thomas, Johannes: thank you for a great time!

I would particularly like to thank Déborah for following me in Germany, and for all these years of patience waiting for me to be done. Merci ma chérie.

Finalement, j'aimerais embrasser ma sœur Anna et sa famille, et remercier du fond du cœur mes parents pour le soutien et l'éducation qu'ils m'ont apportés. Au final, elles auront servi ses heures à réviser des Wörterkasten. Ca y est, on peut dire qu'elle est finie la thèse !

Abstract

Materials in many modern small-scale applications are under complex cyclic stress states and undergo up to 10^9 cycles. Fatigue mechanisms limit their lifetime and lead to failure. Therefore, the Very High Cycle Fatigue (VHCF) regime needs to be studied. This thesis investigates the fatigue mechanisms and crack initiation of nickel, aluminum and copper on a small-scale in the VHCF regime by means of innovative fatigue experimentation. Firstly, the development and implementation of a novel custom-built resonant fatigue setup showed that the resonant frequency of bending micro-samples changes with increasing cycle number due to the accumulating fatigue damage. Then, additional insights on early damage formation have been explored. Mechanisms, prior to crack initiation, such as slip band formation at a state where it appears in only a few grains, have been observed. Cyclic hardening, vacancy formation and oxidation formation may be considered as possible explanations for early fatigue mechanisms. In addition, the new experimental setup can be used to define parameters needed for crack initiation models. Finally, these crack initiation processes have been experimentally examined for pure aluminum and pure copper.

Kurzfassung

Materialien in modernen Kleinbauteilen stehen häufig unter komplexen zyklischen Spannungszuständen. In der Anwendung werden diese bis zu 10^9 Belastungszyklen ausgesetzt. Ermüdungsmechanismen begrenzen ihre Lebensdauer und führen zum Versagen. Daher ist es erforderlich den Very High Cycle Fatigue (VHCF) Bereich zu untersuchen. Der Fokus dieser Arbeit liegt auf den Ermüdungsmechanismen und der Rissinitiierung von Nickel, Aluminium und Kupfer im VHCF-Bereich bei kleinen Proben. Hierfür wurden innovative Ermüdungsexperimente durchgeführt. Das Aufbauen und die Implementierung eines neuen speziell angefertigten Resonanz-Ermüdungs-Versuchsaufbaus zeigte eine Veränderung der Resonanzfrequenz von Mikrobiegeproben mit zunehmender Zyklenzahl, welche sich auf eine Anhäufung von Ermüdungsschädigungen zurückführen lässt. Im Folgenden wurden weitere Anzeichen einer frühen Schädigungsentstehung untersucht. Mechanismen wie Gleitbandbildung konnten in einem Stadium beobachtet werden, in dem lediglich einige wenige Körner betroffen waren. Zyklische Verfestigung, Lehrstellen- und Oxidbildung können als mögliche Erklärungen für frühe Ermüdungsmechanismen herangezogen werden. Des Weiteren kann der neue Versuchsaufbau dazu genutzt werden Parameter zu definieren, die für Rissentstehungsmodelle benötigt werden. Abschließend wurden die Rissentstehungsprozesse von reinem Aluminium und reinem Kupfer experimentell untersucht.

Résumé

De nos jours, les matériaux sont exposés à des états de contrainte complexes et cycliques dans de nombreuses applications à l'échelle microscopique et peuvent subir jusqu'à 10^9 cycles. Différents mécanismes de fatigue limitent leur durée de vie et conduisent à leur rupture. Le Very High Cycle Fatigue (VHCF) régime doit par conséquent être étudié. Cette thèse se concentre sur les mécanismes de fatigue et l'amorçage de fissures pour du nickel, de l'aluminium et du cuivre à l'échelle microscopique, et ce jusque dans le VHCF régime, par le biais d'expérimentations innovantes en fatigue. Tout d'abord, le développement et la mise en œuvre d'un nouveau banc d'essai sur mesure, utilisant la fatigue en résonance, a montré que la fréquence de résonance en flexion de micro-échantillons évolue avec l'augmentation du nombre de cycles, et ceci en raison de l'accumulation d'endommagement par fatigue. Ensuite, une recherche complémentaire sur la formation précoce de dommages a été conduite. Les mécanismes précédant l'amorçage de fissures, tels que la formation de bandes de glissement alors qu'elle apparaît seulement dans quelques grains, a été observé. Le durcissement cyclique, la formation de lacunes cristallographiques et la formation d'oxydation peuvent être considérés comme des explications possibles de mécanismes précoces en fatigue. Ce nouveau dispositif expérimental peut en outre être utilisé pour définir les paramètres nécessaires aux modèles d'amorçage de fissures. Finalement, ces processus d'amorçage de fissures ont été examinés expérimentalement avec de l'aluminium pur et du cuivre pur.

Symbols and abbreviations

AFM	Atomic Force Microscopy
bcc	body-centered cubic
CAD	Computer-Aided Design
DDD	Discrete Dislocation Dynamics
DFG	Deutsche Forschungsgemeinschaft
EBSD	Electron Backscatter Diffraction
EDM	Electrical Discharge Machining
EELS	Electron Energy Loss Spectroscopy
ETD	Everhart-Thornley Detector
fcc	face-centered cubic
FEM	Finite Element Method
FIB	Focused Ion Beam
FPGA	Field Programmable Gate Array
HCF	High Cycle Fatigue
Hz	Hertz
IFMA	French Institute of Advanced Mechanics
kHz	kilohertz
GHz	Gigahertz
KIT	Karlsruhe Institute of Technology
LCF	Low Cycle Fatigue
MEMS	Micro Electro Mechanical System
MHz	Megahertz

OFHC	Oxygen-Free High Thermal Conductivity
PID	Proportional-Integral-Derivative
PSB	Persistent Slip Bands
PSD	Position Sensitive Device
SEM	Scanning Electron Microscope
SPP	Schwerpunkt Programm
TEM	Transmission Electron Microscopy
TL	Tyreus-Luyben
TLD	Through Lens Detector
UHCF	Ultra High Cycle Fatigue
USA	United States of America
VHCF	Very High Cycle Fatigue
ZN	Ziegler-Nichols

1 Introduction

Mechanical structures are often subjected to complex cyclic loading that can lead to fatigue failure of the construction material. The study of the fatigue behavior of materials is conventionally divided into two areas: Low Cycle Fatigue (LCF) for lifetimes smaller than 10^4 cycles and High Cycle Fatigue (HCF) for higher cycle counts. Fatigue data is usually plotted in a diagram, relating the stress amplitude to the number of cycles to failure. This curve is called the S-N curve or the Wöhler diagram. Technically, a stress amplitude can be defined for which the specimens do not fail up to e.g. $N_f = 10^7$ cycles. This limit is called the endurance limit. It is assumed for body-centered cubic (bcc) metals, that for stress amplitudes smaller than this fatigue limit, the specimen or the structure will have an infinite lifetime. On the contrary, pure face-centered cubic (fcc) metals do not show this fatigue limit in higher cycle numbers, and should be investigated further.

Since many mechanical systems operate at high frequencies, they need to survive very high numbers of loading cycles. Applications in which this topic is of significance are for example: the railway and aerospace industries, gas pipelines, nuclear reactor vessels, and microprocessors and Micro Electro Mechanical System (MEMS). In these applications, the conventional fatigue limit cannot be used to predict the allowable stress amplitude. The stress amplitude applicable to such devices needs to be evaluated at these high cycle numbers since the damaging process might differ considerably. It is therefore understandable that it is crucial to know how to predict the lifetime of materials and structures for cycle numbers above 10^8 in fatigue. A new fatigue research classification, called the Very High Cycle Fatigue (VHCF) regime, was therefore established in the 80s.

This thesis was part of the Deutsche Forschungsgemeinschaft (DFG; English: German Research Foundation) priority program SPP1466 'Unendliche Lebensdauer für zyklisch beanspruchte Hochleistungswerkstoffe'. It aims to study the 'infinite lifetime for cyclically stressed high-performance materials'.

The priority program has the ambition to investigate the fatigue mechanisms involved in the VHCF regime above 10^8 cycles. Three groups of materials of particular importance are being investigated: metallic construction materials, materials for MEMS, and continuous fiber reinforced polymers. The work reported in this thesis relates to the project 'Ermüdungsmechanismen von Al und Cu(-Legierungen) in kleinen Volumina im VHCF-Bereich mittels Simulation und neuartiger uni- und multiaxialer Mikro-Ermüdungsversuche'. It focuses on the fatigue mechanisms of Al and Cu at small-scale in the VHCF regime by means of simulation and novel uniaxial and multiaxial micro resonant fatigue experimentation. Dr. Daniel Weygand and his group took care of simulations using Discrete Dislocation Dynamics (DDD), while Prof. Christoph Eberl and the author worked on the experimental portion.

The objective of this thesis is to develop an innovative experimental method to investigate the fatigue mechanisms, the precursors of crack initiation, in the case of small-scale fcc metals in the VHCF regime. Therefore, a novel custom-built resonant fatigue setup is presented in chapter 1, implemented in section 5.1 and discussed in section 6.1.

Additional insights on early damage formation have also been explored by use of this setup. Mechanisms prior to crack initiation are of high interest. They comprise e.g. plasticity induced surface roughening and slip band formation at a state where it appears in only a few grains. For this purpose, the accurate tracking of the fatigue damage evolution in pure nickel, and pure copper have been observed in section 5.2, and analyzed in section 6.2.

Finally, crack initiation plays a particularly important role in the HCF and VHCF regimes, where it very strongly determines the material lifetime. This thesis reports an innovative way of experimentally examining and modeling the crack initiation processes. Thus, VHCF experiments of pure aluminum and pure copper have been carried out in section 5.3, and compared with literature in section 6.3.

2 Literature review – Fatigue of metals

Fatigue is a process (series of mechanisms) which under the action of stresses or deformations (which may vary over time) changes the local properties of a material. This can lead to the formation of cracks and eventually fracture of the structure. Fatigue is particularly characterized by cyclic stress variations with amplitudes well below the yield strength of the material. The main stages of the fatigue failure of a sample are crack initiation, crack propagation and final fracture.

The parameters strongly influencing the fatigue behavior, and thus the number of cycles to failure of a structure, are the stress or strain amplitude, the average stress, the surface condition and environment wherein the structure is used.

Although nowadays the study of fatigue can be based on theoretical considerations (in particular for fracture mechanics), the characterization of a material, of a part, of a structure, still also requires extensive testing and measurements.

In this chapter, the fatigue history and behavior in general as well as for small-scale samples is introduced. Possible damage mechanisms and models are presented. And finally, the VHCF phenomena and the different testing methods are described.

2.1 History of fatigue in metallic materials

The first listed fatigue tests were conducted by Wilhelm Albert (German mining administrator) in 1829 and concerned mine shaft chains [1]. In 1839, Jean-Victor Poncelet (French engineer and mathematician) used for the first time the term 'fatigue'.

In 1842, the fatigue failure of an axle causes the Versailles rail accident, as has been shown in the works of William Rankine [2]. In 1852, August Wöhler (German engineer) is charged by the Prussian Minister of Commerce to work on ruptures of train axles [3]. He exhibited his work at the World Expo 1867 in Paris.

More recently, in 1954, Coffin and Manson worked on fatigue and explained cyclic crack growth based on plastic strains at the crack tip [4, 5]. They showed that the crack initiation in ductile metals results from plastic deformation localized on the surface of the specimen. The accumulation of cycles results in plastic deformation at the macroscopic level resulting in damage formation and failure.

Nowadays the research on fatigue is common and well developed. For example Suresh reviewed it in details in his book entitled "Fatigue of materials" [6]. Mughrabi also reviewed the fatigue of metals, especially with respect to very high cycle fatigue, as well as fatigue of ultrafine grain or nanocrystalline metals [7], or on crack initiation caused by fatigue irreversibilities [8].

2.2 S-N curve – Fatigue lifetime characterization

The fatigue lifetime is often represented by the Wöhler curve, also called S-N curve. It defines a relationship between the applied stress amplitude $\sigma_a=\Delta\sigma/2$ (sometimes called S) and the number of cycles to failure N_f. **Fig. 2.1** is a typical S-N curve, which represents the fatigue life characterized by four stages [9]. One can distinguish:

I. The Low Cycle Fatigue regime for which the specimen failure occurs at a small number of cycles (less than 10^4-10^5 cycles). Strain amplitudes are around and above the macroscopic elasticity limit of the material.

II. The High Cycle Fatigue regime for which the specimen failure occurs at a number of cycles between 10^5 and 10^7. In this case, the stress amplitudes are at least two times lower than the macroscopic elasticity limit of the material. The S-N curve tends to a horizontal asymptote representing the endurance limit of the material. This asymptote is more or less well defined depending on the materials. It also defines a conventional fatigue limit which corresponds to the resistance to fatigue of the material at 10^7 cycles. Below this endurance limit, the lifetime of the material has long been considered as infinite and is therefore used for designing mechanical fatigue parts.

III. The Very High Cycle Fatigue regime, also sometimes called Ultra High Cycle Fatigue (UHCF) by the German community or gigacyclic fatigue by the French community. Strain amplitudes are less than the conventional fatigue limit. However, the material may fail after a number of cycles greater than 10^7 cycles.

IV. The question of the existence of a fatigue limit remains today an open question.

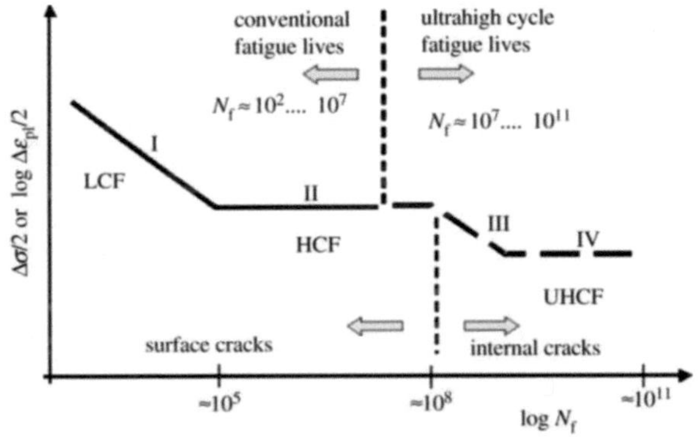

Fig. 2.1: Wöhler curve (or S-N curve) - with the number of cycles to failure N_f depending on the amplitude of the applied stress – from [9].

The S-N curve in **Fig. 2.1** is a schematic curve. The distinction between the different regimes is much less pronounced in reality. Sometimes, the presence of a plateau in the HCF regime (regime associated with the conventional fatigue limit) is not even observed (**Fig. 2.2**).

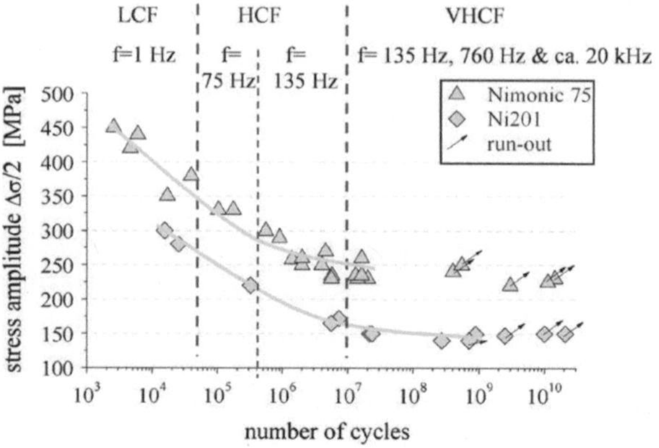

Fig. 2.2: Wöhler curve (or S-N curve) for Nimonic 75 and pure nickel (arrows indicate run-out specimens) in the LCF–VHCF regime – from [10].

2.3 Fatigue damage formation from dislocations to micro-cracks

As already indicated for ductile materials; the crack initiation results from plastic deformation, localized close to the specimen surface.

2.3.1 Dislocations

The plastic deformation results from glide along well-defined parallel crystallographic planes (**Fig. 2.3**). This gliding occurs on crystallographic planes of high density, called glide planes, and follow particular directions within those planes.

These directions are the dense directions of the glide plane; it means the directions in which the distance between atoms is the smallest. They are called glide directions. The combination of a glide plane and a glide direction forms a glide system. In the fcc crystal structures, such as Ni, Al or Cu, there are twelve possible glide systems. There are four glide planes of the {111} family in which three possible glide directions are of the <110> family. Glide systems are activated (the planes glide) when the shear stress on a glide plane and in the glide direction reaches a critical value, called the critical shear stress. The activation of glide systems not only produces a plastic deformation but also a rotation of the crystal lattice to satisfy the motion continuity.

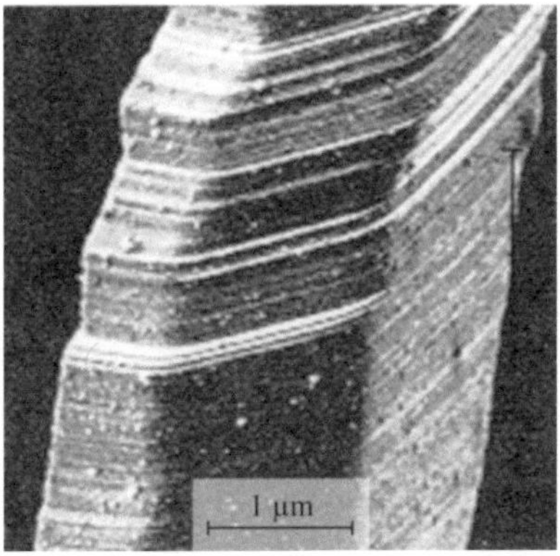

Fig. 2.3: Scanning electron micrograph of a hexagonal cobalt single crystal showing the surface features associated with glide planes – from [11].

The glide along a plane results from the motion of linear lattice defects referred to as 'dislocations'. There are two types of dislocations: edge dislocations where the Burgers vector is perpendicular to the dislocation line and screw dislocations where the Burgers vector is parallel to the dislocation line (**Fig. 2.4**). The dislocation density is the total length of dislocations per unit volume (m/m^3). It increases with plastic deformation due to the creation of new dislocations.

Dislocations can also be eliminated at interfaces or by annihilation of dislocations with opposite Burgers verctor, for instance, during an appropriate heat treatment (annealing).

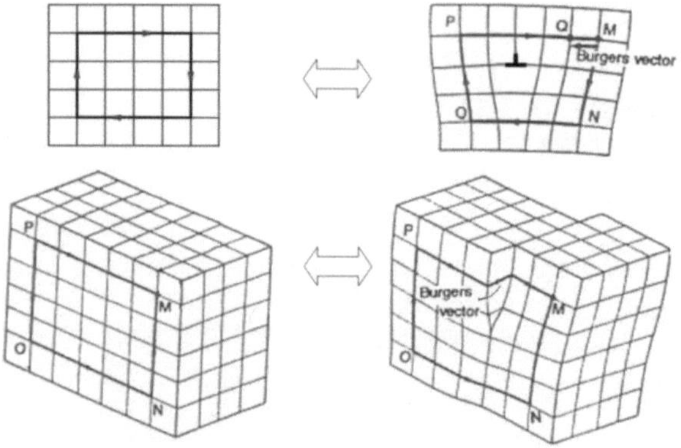

Fig. 2.4: Top right: schematics of an edge dislocation. Bottom right: screw dislocation. Source: https://en.wikipedia.org/wiki/Dislocation.

The glide plane of an edge dislocation is unique and defined by the direction of the dislocation line and the direction of the Burgers vector. On the contrary, the screw dislocation has several glide planes since its dislocation line and its Burgers vector are parallel. During deformation, it can change the glide plane, which is called cross slip. The primary slip system (first dislocation glide plane) and the deviated system have the same direction but different glide planes. This possibility exists only for screw dislocations.

2.3.2 Slip bands

The appearance of slip lines and associated dislocation structures depend on many factors such as the loading mode (monotonic or cyclic, uniaxial or multiaxial ...), the crystallographic structure, the crystallographic orientation with respect to the mechanical loads, temperature, the strain rate, etc.

Nevertheless, observations of the surface of a plastically deformed metallic sample show the presence of slip lines, often gathered in bands called slip bands. They look like steps and are created by the emergence of dislocations on the sample surface.

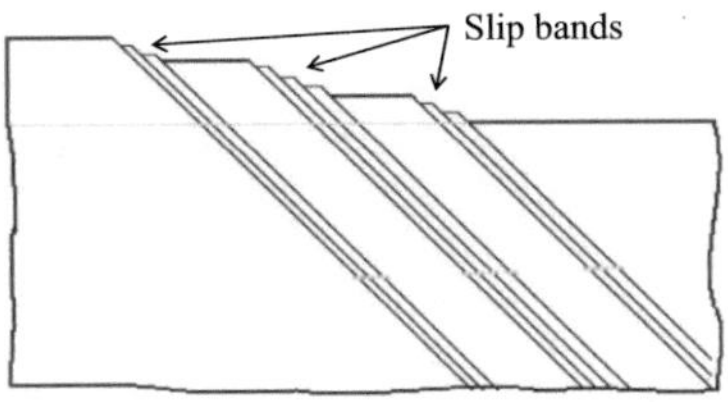

Fig. 2.5: Schematics of a crystal surface having several slip lines gathered in slip bands.

2.3.3 Crack initiation

Cracks initiate on the sample surface:

I. at the location of intense slip bands which emerge from the specimen surface, they can also be called Persistent Slip Bands (PSB). In this case, this is an intragranular (within a grain) crack formation, also called transgranular crack formation,

II. at grain boundaries resulting in an intergranular (between grains) crack formation.

This is detailed in the following section:

I. In literature, the research mostly focused on the formation of cracks initiated at PSBs for two reasons: First of all, this mode of failure is common and even more frequent when the constraints are sufficient to enable plasticity in grains. Secondly, many studies focus on single crystals and thus without grain boundary.

The crack initiation at a PSB occurs at the interface between the PSB and the matrix. **Fig. 2.6** shows a crack initiated along a PSB in a copper single crystal [12]. The observation also revealed that the deformation in the PSB is strongly inhomogeneous.

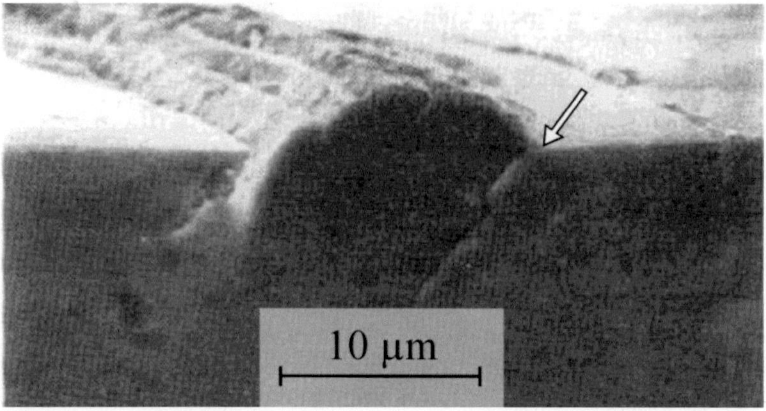

Fig. 2.6: Scanning electron micrograph of the initiation of a fatigue crack (arrow) at the border of a PSB in copper – from [12].

II. The interaction between a PSB and a grain boundary can also be the origin of the crack initiation at grain boundaries [8, 13]. In an optical interferometric measurement of slip band heights at fatigued Cu grain facet, Kim and Laird [14] noted that the fatigue crack may occur at grain boundaries if:

- The misorientation between the two grains separated by the grain boundary is high,
- the glide plane activated within at least one of the two grains has a slip band at the specimen surface perpendicular to the grain boundary,
- the angle between the loading axis and the grain boundary is in the range of 30-90°.

Fig. 2.7 shows a fatigue crack nucleation at the grain boundary for a polycrystalline copper. In this case, the PSBs form a non-zero angle with the grain boundary. Nevertheless, another type of the PSB could grow along the grain boundary.

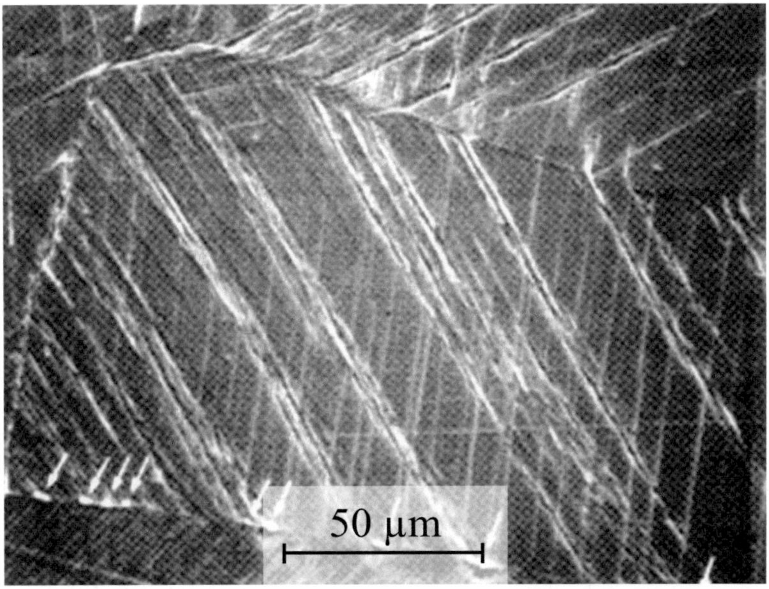

Fig. 2.7: Scanning electron micrograph of the crack initiation (emphasized by arrows) at the grain boundary in copper – from [13].

In conclusion, regardless of the initiation site of the crack, the formation of PSBs is the precursor mechanism of the crack initiation. Thus, the conventional fatigue limit is associated with the stress (plastic deformation) required to form PSBs. As an example, for polycrystalline copper under high stress amplitudes, cracks initiate at the grain boundary, whereas at low stress amplitudes, they may initiate at PSBs but also at the grain boundary.

2.3.4 Role of fatigue-induced surface oxides

As previously shown, the phenomenon of slip band formation corresponds to an extrusion/intrusion formation mechanism. The dislocation motion happening with the slip band formation can also lead to surface oxidation processes [6, 15].

This kind of cyclically induced slip band motion implies the appearance of freshly exposed metal, which will oxidize after each fatigue cycle (**Fig. 2.8** shows PSB-induced oxidation mechanism [16]).

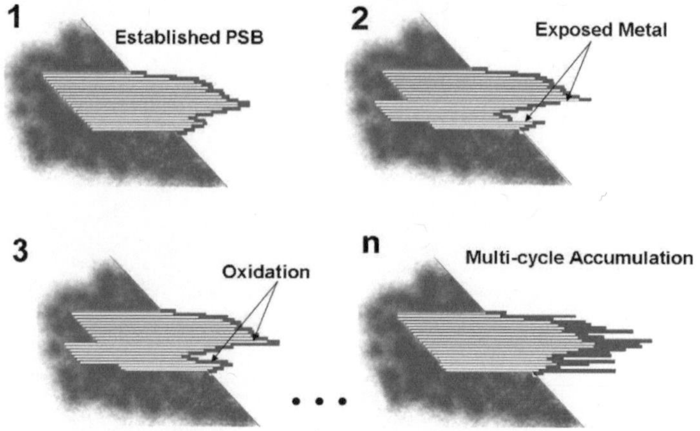

Fig. 2.8: Schematics of a PSB-induced oxidation mechanism – from [16].

Thompson et al. [17], and more recently Boyce et al. [18] have suggested the possibility that oxygen chemisorbed on the free surface of a slip band could be incorporated into the sample when the load reversal causes the exposed, oxidized surface to slip back into it. Boyce's observations revealed a new level of understanding of the role of oxygen in PSB crack initiation. They stated that atmospheric oxygen serves to form an oxide on the surface of the metal, which is self-passivating prior to fatigue loading.

However, the motion of PSB extrusions and intrusions on the surface (see Scanning Electron Microscope (SEM) image of the sample surface **Fig. 2.9**a) results in a disruption of the passivated oxide, allowing for new diffusion paths for atmospheric oxygen to react with the underlying base metal, thereby resulting in an oxide thickening mechanism.

In electrodeposited nickel, at fatigue stresses slightly above the endurance limit, this surface oxide (see Electron Energy Loss Spectroscopy (EELS) of a Focus Ion Beam cross section **Fig. 2.9**b and c) was found to be as thick as 400 nm (**Fig. 2.9**d), several orders of magnitude thicker than the native oxide. The role of oxygen on fatigue life has been established for many decades; Boyce's study simply asserts that the oxygen is forming a thick brittle oxide on the PSB surface. Further investigation is necessary to determine if the currently observed mechanism extends to other ductile metals, including Cu which was used as a model material in studies (e.g. the work of Ma [12]) on the role of PSBs in crack initiation.

Baumert et al. [19] also observed the critical role of highly localized oxide thickening on the fatigue crack nucleation process in electroplated Ni MEMS specimen in the VHCF regime at large stress amplitudes under extreme stress gradients. Transmission Electron Microscopy (TEM) revealed localized surface oxides as thick as 700 nm. These observations are consistent with the "PSB oxide thickening" mechanism described in [18], although here, the localized oxides likely form at the location of cyclic slip bands instead of PSBs. Microcracks were observed inside locally grown oxides, which likely form at the location of cyclic slip bands (**Fig. 2.10**).

Finally Shan et al. [20] observed, after fatigue loading until sample failure, a nickel oxide thickness increased from 4 nm to 9 nm during a fatigue life of 1.7×10^5 cycles. In addition, Atomic Force Microscopy (AFM) analyses revealed the presence of rough surfaces with topography variations in the micron-scale range. The observed increase in oxide film thickness is consistent with prior work of Boyce et al. [18], whose samples were also fabricated at the Sandia National Labs following the same procedure as Shan. While Boyce emphasized local increase of oxide thickness on the order of a few hundred nanometers, Shan's study evidenced an increase in the oxide thickness on average.

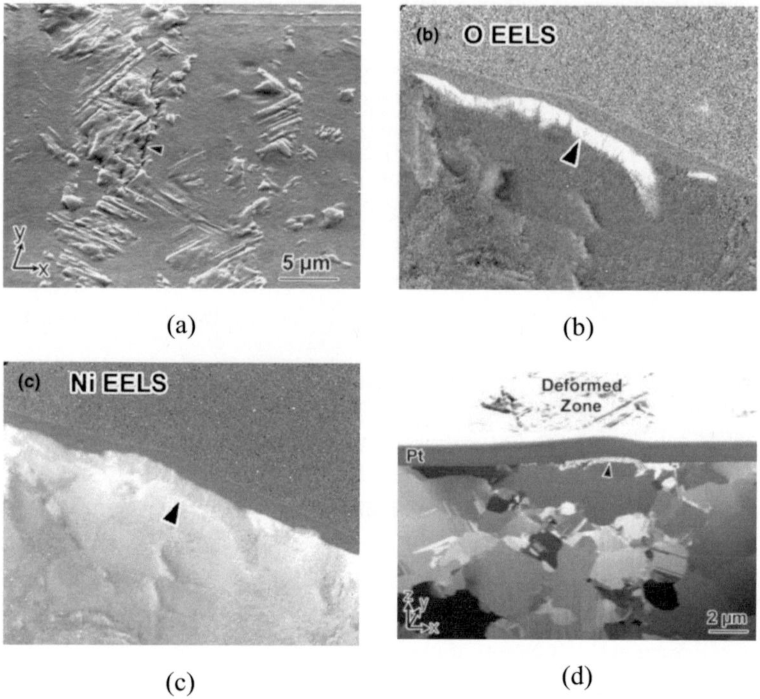

Fig. 2.9: (a) Scanning electron micrograph of the localized zones of deformation found on the sidewall surfaces of a Ni-sample where the bending stresses were maximal. The arrow indicates a micro-crack that has formed along the edge of one deformed zone. (b) Spectral correlation map of the presence of oxygen (bright) as detected by EELS, showing a large concentration of oxygen present in the surface layer. (c) A similarly constructed nickel map indicates the presence of nickel in the bulk and the surface layer. (d) Ion-induced scanning electron micrograph revealing the subsurface cross-section of a deformation zone exposed during focused ion beam milling. A Pt layer had been deposited on the surface for protection during ion milling. The arrow identifies the highly contrasted region of interest at the surface of the grain associated with the deformation zone – from [18].

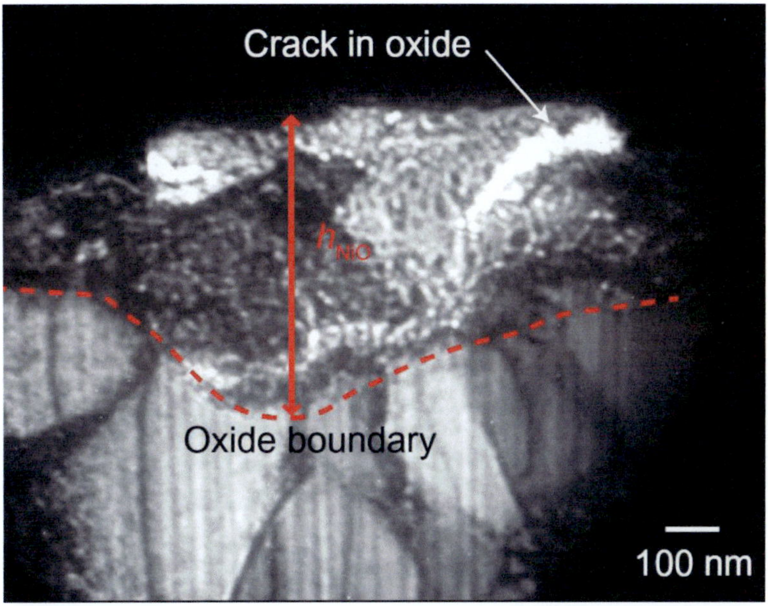

Fig. 2.10: TEM images of a fatigued notch showing highly localized formation of surface oxide, at the location of extrusions. h_{NiO} corresponds to the thickness of the oxidation layer – from [19].

2.4 Fatigue models

This section is based on the detailed review of Chan [21] on the status of microstructure based fatigue crack initiation models since the 80's, up to today (2015). According to Cowles [22], experimental observations and lifetime prediction systems can be expensive to establish, due to the need of large experimental databases. And this is why, in a cheap and fast way, lifetime prediction model may improve the understanding of fatigue crack initiation.

2.4.1 Empirical lifetime models

For uniaxial cyclic loading in metals, the first fatigue crack initiation models have been developed in the 80's. For example in terms of the stress-life approach by Mitchell [23]:

$$\sigma_a = \sigma_f' N_f^\alpha \qquad \text{Eq. 2.1}$$

Where σ_a is the stress amplitude, σ_f' is the fatigue-strength coefficient, N_f is the number of cycles to failure, and α is the fatigue-strength exponent, varying between 0.5 and 1.

In terms of the strain-life approach, the plastic strain range ($\Delta\varepsilon_p$) is related to the cycles to failure according to the Coffin-Manson relation [4, 5]:

$$\frac{\Delta\varepsilon_p}{2} = \varepsilon_f' N_f^b \qquad \text{Eq. 2.2}$$

Where ε_f' is the fatigue-ductility coefficient and b is the fatigue ductility exponent.

These two models are nowadays well known, however they do not take into account the specimen microstructure in the crack initiation process or the lifetime prediction.

2.4.2 Crack initiation models

Yet, as shown in 2.3.3, the crack initiation mechanisms are associated with the formation of PSBs. This has been implemented by Cheng and Laird [24] by adding in Eq. 2.2 a crack nucleation criterion based on the formation of a critical notch depth:

$$\frac{\Delta\gamma_p}{2} N_i^\alpha = C' \qquad \text{Eq. 2.3}$$

Where $\Delta\gamma_p$ is the plastic shear strain range, N_i the number of cycles to crack initiation, α is the fatigue-strength exponent, varying between 0.5 and 1 (as in Eq. 2.1) and C' a constant.

At the same time, Tanaka and Mura proposed a fatigue crack initiation model based on the accumulation of dislocation dipoles at a slip band [25] during strain cycling:

$$(\Delta\tau - 2k)N_i^{1/2} = \left[\frac{4\mu W_s}{\pi(1-\nu)d}\right]^{1/2} \qquad \textit{Eq. 2.4}$$

Where $\Delta\tau$ is the shear-stress range, k is the friction stress of a dislocation, μ is the shear modulus, d is the grain size, v is the Poisson's ratio and Ws is the specific fracture energy per unit area along the slip band.

The same model was proposed for crack initiation at the grain boundary, in this case W_S is replaced by W_G the specific fracture energy per unit area along the grain boundary. On the same principle, a fatigue crack initiation model based on inclusion fracture because of impinging slip and dislocation pileup [26] was proposed:

$$(\Delta\tau - 2k)N_i^{1/2} = \left[\frac{4\mu(\mu+\mu')W_I}{\pi\mu'\xi}\right]^{1/2} \qquad \textit{Eq. 2.5}$$

Where μ' is the shear modulus of the inclusion, ξ is the inclusion size, and W_I is the specific fracture energy per unit area of inclusion.

2.4.3 Stress intensity threshold model for crack initiation

A different approach has been proposed by Harvey et al. [27] and depends on the deepening of slip steps into a fatigue crack during strain cycling. AFM has been used for measuring the slip height displacements (couple of nm). And they argued that the crack initiation is determined by the stress intensity threshold ΔK_{th}:

$$N_i = \frac{\Delta K_{th}^2}{4\sigma_{ys}Ef\Delta\varepsilon_p h} \qquad \textit{Eq. 2.6}$$

Where E is the Young's modulus, σ_{ys} is the yield stress, $\Delta\varepsilon_p$ is the plastic-strain range, f is the fraction of plastic strain range that produces the surface displacement, and h is the slip band width.

Nevertheless, none of these fatigue crack initiation models contain a crack length parameter, and as a consequence cannot be used for a prediction of the size of a fatigue crack at initiation. Chan [21] specifies the difficulties to detect cracks smaller than 100 nm with AFM. And therefore, the need exists for a methodology that can predict the number of cycles to initiation for a fatigue crack of a given size.

2.4.4 Crack initiation model with crack length parameter

The following model was proposed by Chan [21] and is an extension of the dislocation dipole model of Tanaka and Mura [25] taking into account a crack size based model. First of all Tanaka and Mura [25] showed that the change in the stored strain energy (ΔU) of dislocations per unit thickness between the forward and reverse fatigue loading is:

$$\Delta U = \frac{\pi(1-\nu)(\Delta\tau - 2k)^2 d^2}{4\mu} \qquad \textit{Eq. 2.7}$$

They then showed that the number of fatigue cycles to crack initiation was obtained by equating the stored energy in the dislocation dipoles to the specific fracture energy (W_s) per unit area. The energy balance is then equal to:

$$2N_i \Delta U = 2dW_s \qquad \textit{Eq. 2.8}$$

Which finally corresponds to Eq. 2.4 and gives:

$$(\Delta\tau - 2k)^2 N_i = \frac{4\mu W_s}{\pi(1-\nu)d} \qquad \textit{Eq. 2.9}$$

This shows the inverse proportionality between fatigue crack initiation life and the grain size d. This model is still nowadays used for prediction of crack initiation at PSBs, and can be used for example in the VHCF regime.

Though, the crack size is arbitrary and cannot be predicted. A last extension done by Mura and Nakasone [28] can be conducted where the crack size is implemented by considering the Gibbs free energy change ΔG associated with the nucleation of a fatigue crack from a double pileup of dislocation dipoles:

$$\Delta G = -W_e - W_m + 2a\gamma_s \qquad \textit{Eq. 2.10}$$

W_e is the elastic strain energy per unit thickness stored in dislocations, which is comprised of the self-energy and the interaction energy of the dislocation dipoles. The value of W_e is given by:

$$W_e = \frac{\xi_1 d^2(1-\nu)(\Delta\tau - 2k)^2 N^2}{2\pi\mu} = z_1 N^2 \qquad \textit{Eq. 2.11}$$

Where ξ_1 is a numerical constant.

The W_m term is Irwin's strain-energy release rate per unit thickness due to the crack opening and is given by [28]:

$$W_m = \frac{\pi d^2(1-\nu)^3(\Delta\tau - 2k)^2 \Delta\tau^2 N^2}{2\mu^3} = z_2 N^2 \qquad \textit{Eq. 2.12}$$

Furthermore, the incipient crack length can be related to the dislocation pileup, and described as:

$$a = \frac{2d(1-\nu)(\Delta\tau - 2k)N}{\mu} = z_3 N \qquad \textit{Eq. 2.13}$$

Finally, the Gibbs free energy change from Eq. 2.10 can be simplified:

$$\Delta G = -z_1 N^2 - z_2 N^2 + 2 z_3 N \gamma_s \qquad \textit{Eq. 2.14}$$

where γ_s is the surface energy (per unit area) of the crack.

The free-energy change can described as a function of fatigue cycles. Initially, the free-energy change is positive ($\Delta G > 0$), indicating that there is an energy barrier for crack initiation. The free-energy change reaches a maximum where $\partial \Delta G / \partial N = 0$, which indicates that the dislocation-dipole structure and the incipient crack are energetically in quasi equilibrium. Beyond the maximum, $\partial \Delta G / \partial N < 0$, the dislocation dipole structure is unstable with respect to crack formation.

They [28] also based their research on the fact that the number of dislocation dipoles in a single slip band is equal to:

$$n_{eq} = 0.05\left(\frac{d}{bh}\right)\sqrt{\frac{W_{eq}}{\mu}} \qquad \textit{Eq. 2.15}$$

Where h is the slip band width and W_{eq} is the strain energy per unit thickness stored in the dislocation dipoles of a single or an equivalent slip band.

Nevertheless, Mura and Nakasone [28] argue that not all dislocation dipoles contribute to the formation of the fatigue crack, at least not immediately at crack initiation. Under this circumstance, it is more appropriate to consider fatigue-crack initiation as occurring only at a portion of the slip band. The surface energy required to create this crack would be $2a\gamma_s$, and the strain energy within the incipient crack that would be released to form the crack is aW_{eq}/d. The maximum in the Gibbs free energy can be postulated to occur when the following condition is satisfied:

$$2a\gamma_s = \frac{aW_{eq}}{d} \qquad \textit{Eq. 2.16}$$

This leads to:

$$W_{eq} = 2d\gamma_s \qquad \textit{Eq. 2.17}$$

Knowing that the crack length a at crack initiation is given by the number of dislocations n_a:

$$a = n_a b \qquad \textit{Eq. 2.18}$$

Where b is the magnitude of the Burgers vector.

The value of n_a is obtained from Eq. 2.15 by replacing W_{eq} by aW_{eq}/d, since only that amount of strain-energy density is required to overcome the surface energy.

$$n_a = 0.05\left(\frac{d}{bh}\right)\sqrt{\frac{aW_{eq}}{d\mu}} \qquad \textit{Eq. 2.19}$$

It can be concluded by combining it with Eq. 2.17 and Eq. 2.18 that the crack length $2a$ or depth a is equal to:

$$a = 0.005 \left(\frac{d}{h}\right)^2 \left(\frac{\gamma_s}{\mu}\right) \qquad \textit{Eq. 2.20}$$

And finally, using Eq. 2.9 and imposing the condition that $\gamma_s = W_s$, the criterion for the initiation of a fatigue crack in polycrystalline materials is:

$$(\Delta\tau - 2k)N_i^{1/2} = \left[\frac{4\mu^2}{\lambda\pi(1-\nu)}\right]^{1/2} \left(\frac{h}{d}\right)\left(\frac{a}{d}\right)^{1/2} \qquad \textit{Eq. 2.21}$$

Where λ represents the cyclic irreversibility ($0 < \lambda < 1$).

In 2015, Krupp et al. [29] developed a Finite Element Method (FEM) model based on the previous crack initiation models. The finite element modeling concept has been developed further to analyze three-dimensionally the crack initiation process and the crack interaction with microstructural barriers (grain and phase boundaries). Crack initiation and stepwise propagation is treated by using the model of Tanaka and Mura [25] and Chan [21]. As already reviewed in this section, Tanaka and Mura assumed that dislocations pile up at the left and the right grain boundary of a representative grain (dislocation double pile up). Crack initiation is accounted for by an energy balance. Once the stored energy by the accumulated dislocation pile up equals the surface energy, the dislocation dipoles transform in free surface (=crack). This condition for the initiation of a crack of length a at a number of cycles N_i can be written according to Chan [21] (equivalent to Eq. 2.21) as:

$$N_i = \frac{4\mu^2 h^2 a}{\lambda\pi(1-\nu)(\Delta\tau - 2k)^2 d^3} \qquad \textit{Eq. 2.22}$$

2.5 Very High Cycle Fatigue (VHCF)

Mughrabi [9] proposed to classify materials into two types according to their crack initiation mechanisms in the VHCF regime.

Type I materials: These are single phase ductile metals or alloys which have neither precipitate nor inclusion. Their microstructure is simple. This is the case, e.g., of pure metals with an fcc structure. Observations show that the cracks initiate on the surface of the specimen.

Type II materials: These are metals and alloys with a more complicated microstructure, multi-phase or with precipitates. High strength steels are typical of type II materials. Observations show that cracks are initiated at subsurface inclusions in the specimen volume.

The work presented in this thesis is a study of type I materials, such as Ni, Al, and Cu, with an fcc crystal structure. In this type of material, two obstacles for the dislocation motion are the grain boundaries and other dislocations.

Studies of fatigue in the LCF and HCF regimes are often conducted using a servo-hydraulic machine where the frequency is less than 200 Hz. In 1950, Mason has developed the first ultrasonic fatigue machine capable of working at 20 kHz frequencies. This first machine was the basis for the design of a large number of ultrasonic machines used today. Ultrasonic fatigue tests are very interesting because they allow for measuring quickly the material behavior under many fatigue cycles and exploring a very extensive range of cycles up to 10^{11} cycles.

Recent experiments on body-centered cubic metals [30] explored the VHCF regime beyond 10^7 cycles and distinguish two regimes e.g. for high strength steels (**Fig. 2.1**). The lower regime corresponds to the transition from the HCF limit (e.g. 10^7 cycles) to the VHCF regime (e.g. 10^9 cycles) [31, 32, 33], and [34]. In this domain, a surface roughening can be observed due to irreversible cyclic slip of dislocations. The fourth regime represents the VHCF limit, below which cyclic slip would become essentially reversible and non-damaging.

For pure fcc metals, it is still unclear if there is a fatigue limit similar to such a regime. Al thin films do not show such a limit for up to 10^{14} cycles for a stress amplitude of roughly 14 MPa [35]. In this case, the experiments were carried out at 1 GHz and the authors argue that at such high frequencies the fatigue mechanisms can differ from those at low frequency fatigue [36].

Nevertheless, it has been shown for pure copper that the lifetime in the VHCF regime is controlled by crack initiation as a consequence of irreversible plastic deformation at the surface, such as slip bands [34, 37, 38, 39], and [40]. Mughrabi [41] proposed that the development of surface roughening is precursor for the crack initiation (**Fig. 2.11**). Even at very low stress amplitudes, crystallographic slip occurs and a part of it is irreversible. The amount of irreversibility is very small and decreases with decreasing the stress amplitude. However, the accumulation of this irreversibility produces a small surface roughness. If a critical state of the surface roughness is reached, the stress concentrations at some slip bands are strong enough to allow for PSB formation and crack initiation. Some cracks can then propagate and cause final failure of the specimen. Fatigue failure can thus occur at stress levels below the conventional fatigue limit after a large number of cycles.

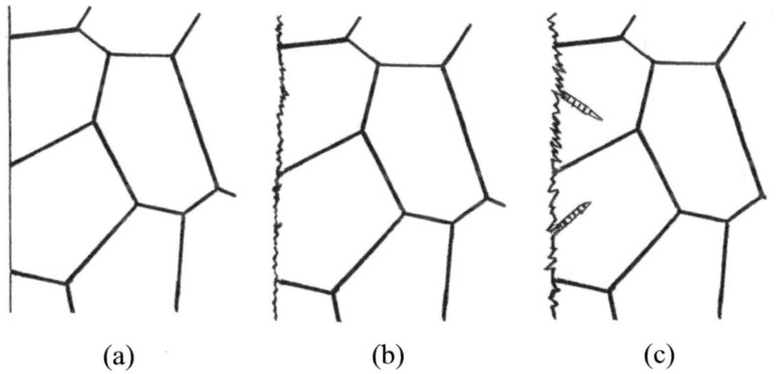

Fig. 2.11: Schematics of fatigue mechanisms for stresses lower than for the 'conventional PSB threshold' (a) initial state, (b) surface roughness formation and (c) PSB formation, followed by crack initiation – from [41].

In conclusion, the slip bands can lead to the initiation of micro-cracks, where one of them may undergo continued short crack growth, followed by fatal failure. In the case of VHCF, the lifetime is mostly determined by crack initiation followed by micro and short crack growth, while the macroscopic crack growth is only a small part of the total lifetime [41]. Therefore, experiments, where the crack initiation can be detected, are important for tracking damage evolution, since grain size and orientation have important implications on the topic [42]. For bulk samples, VHCF experiments are typically carried out at much lower frequencies of 20 kHz with resonant fatigue setups. Commonly such setups induce a fully reversed uniaxial stress state [43]. An ultrasonic fatigue setup for torsional fatigue testing was implemented by the group of Bathias [44].

2.6 Size and scaling effects

The sample size plays an important role on the fatigue lifetime. Furuya showed for high strength steel bulk samples that the mean VHCF lifetime increases by an order of magnitude with decreasing specimen volume from ~800 mm^3 to ~30 mm^3 [45]. This is known as a classical size effect which occurs when the tested samples are too small to represent the microstructural feature and defect distribution. Following a random distribution, the largest defects typically exist only in limited numbers within smaller samples, and more negatively impact the lifetime of the sample compared to smaller defects.As a result, the mean lifetime of smaller samples seems to increase, while generally, for all sample dimensions, the trend is toward an increasing scatter at higher lifetimes.

Small-scale fatigue is also relevant to the applications and research of many current material science and engineering topics, such as thin films [46], MEMS [47], surface acoustic wave devices [36] or microelectronics. Such systems are typically constrained by their complex structure and often work at particularly high frequencies of up to several GHz [35, 48].

Therefore, required lifetimes are in the high (until 10^7) and very high (beyond 10^8) cycle regime, while cyclic loading conditions can be multiaxial leading to damage accumulation until failure.

For very small dimensions as in thin films, MEMS or nanostructured materials, the damage mechanisms can be constrained or even restricted. Here, the extrinsic and intrinsic dimensions approach the size of the damage governing mechanisms and processes, e.g. dislocations and dislocation structures. This has been observed for metal thin films, where film thickness is inversely proportional to fatigue strength [49]. The microstructure of such films is smaller than typical fatigue dislocation structures (e.g. persistent slip bands). Thus, such characteristic dislocation structures cannot develop. Although the size constraining effects on mechanical properties or fatigue behavior are not fully understood, such materials are used in safety relevant systems. Furthermore, little research has been conducted on small-scale multiaxial fatigue which is the typical loading condition in a real environment.

2.7 Experimental methods for crack initiation and propagation observation

Techniques to track crack initiation and propagation are e.g. the replica technique, electrical damage detection, thermal methods, and the analysis of the higher harmonics in sample behavior.

The replica technique is a well-known method for examining the crack profile of macro samples [50]. After being moistened in acetone, a cellulose acetate sheet is pressed against the sample surface. The replica is then peeled off, coated e.g. by Au, and then examined in an SEM. This method allows for post-mortem analyses at different fatigue stages while the sample remains in the testing machine. However this method does not allow for the continuous tracking of crack formation and growth. Furthermore, it is necessary to apply a load to imprint the surface roughness and is thus not well suited to be used on fragile micro-samples.

Another way of tracking crack initiation and propagation is electrical damage detection. The electrical resistance of a sample depends on its cross section, which becomes smaller as cracks propagate. Early electrical resistance changes can be attributed to the initiation of voids, dislocations and micro-cracks. However, it is not always possible to identify micro-cracks shallower than 0.1 mm [51]. Sun et al. showed that fatigue lifetime prediction and failure [52] can be established, but the detection of crack initiation remains difficult. For an experiment operating at the resonant frequency of the sample, it is difficult to attach electrical connections without influencing the vibrational mode.

This can be circumvented by infrared thermography methods [53]. The temperature field on the sample surface is observed by an infrared camera. At the crack initiation site, thermal dissipation leads to a surface temperature increase. That said, the spatial resolution of the infrared camera limits the detection of micro-crack formation. These two methods, infrared thermography and electrical resistance measurement, could also be combined in a new lifetime determination [54].

Another mechanical method, which can be implemented as a non-contact in-situ measurement of non-linear behavior was introduced by Kumar et al. and Funk et al. [55, 56]. Thc authors describe their method as a nonlinear ultrasonic measurement via analysis of the feedback signal of a closed-loop ultrasonic fatigue system, where the amplitude of the higher harmonics of the feedback signal is analyzed. The authors argue that the current damage state leads to a specific non-linear behavior, corresponding to changes in the amplitude of the higher harmonics of the (applied) fatigue frequency. Both internal and surface-initiated cracks, as well as initial hardening/softening, and fatigue crack growth can be detected. So far, the characteristic answer is difficult to trace back to a specific damage mechanism, although this might be possible in the future.

2.8 Summary

The literature review showed strong similarities in the mechanisms of plastic deformation and crack initiation between HCF and VHCF regimes. In both cases, crack initiation is related to the presence of slip bands on the specimen surface, creating a surface roughness increasing with number of cycles. This increasing roughness creates stress concentration sites and leads to crack initiation. However, differences exist. In the HCF regime, the fatigue limit is mostly associated to the PSB formation threshold. In the VHCF regime, for fcc metals, no fatigue limit even at 10^{10} cycles was clearly found. However, slip bands are observed at small stress amplitudes.

Finally, some mechanisms, such as fatigue-induced surface oxides, and some models, Tanaka-Mura, Chan, and finally Krupp investigate the fatigue crack initiation behavior and prediction.

The objective of this thesis is to contribute to a better understanding in the VHCF regime of the damage formation, precursors of the crack initiation, as well as their nature and their formation requirements related to the microstructure for fcc materials. Therefore, a novel custom-built resonant setup has been developed for carrying out fatigue tests. The early damage formation, as well as the crack initiation have been characterized by observation of the sample surface.

3 Custom-built resonant fatigue setup

3.1 Experimental principle

Multiaxial fatigue experiments with micro-samples at kHz frequencies require novel experimental setups. For bulk samples and MEMS devices, such apparatus are typically designed as resonant devices and can only be used at their characteristic frequencies to maximize the stress amplitude. In the here presented novel experimental setup such a principle is implemented as well.

The initial approach is to fix the sample on one side with a clamp and to displace this clamp by two independently controlled piezo actuators normal to the sample surface (P1 and P2). In-phase, a cyclic actuation at P1 and P2 leads to a bending vibration of the sample (**Fig. 3.1**a), while an anti-phase excitation at P1 and P2 leads to a torsional vibration (**Fig. 3.1**c). A combined signal leads to a multiaxial excitation, a combination of bending and torsional oscillations.

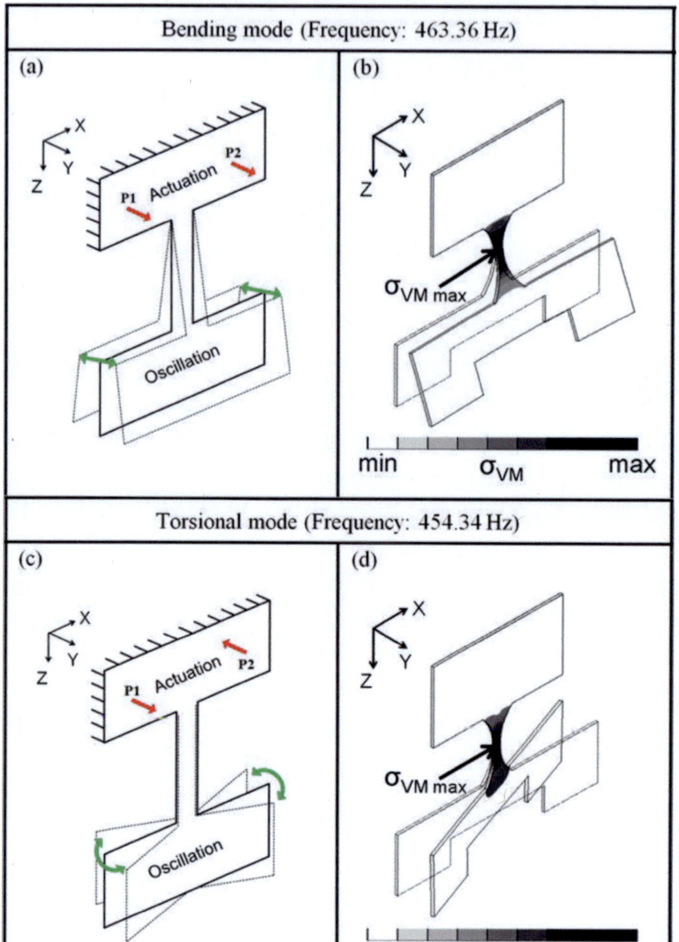

Fig. 3.1: Schematics representing the actuation and oscillation for the bending (a) and the torsional (c) mode. The locations of actuation by the two piezo actuators are marked with arrows P1 and P2. Finite element modal analysis showing the maximal stress in the gage section of the samples for the bending (b) and torsional (d) mode.

3.2 Sample geometry optimization by FE simulation

3.2.1 "T" shape sample geometry parameter pre study

An initial parameter study has been undertaken to observe the geometry influence on the resonant frequencies of the sample. For this study, the sample has a "T" shapc (**Fig. 3.2**). For the purpose of a modal analysis, all the nodes of the bottom area are blocked, the sample is not submitted to a load, and the end of the sample, where the mass is, is free to move.

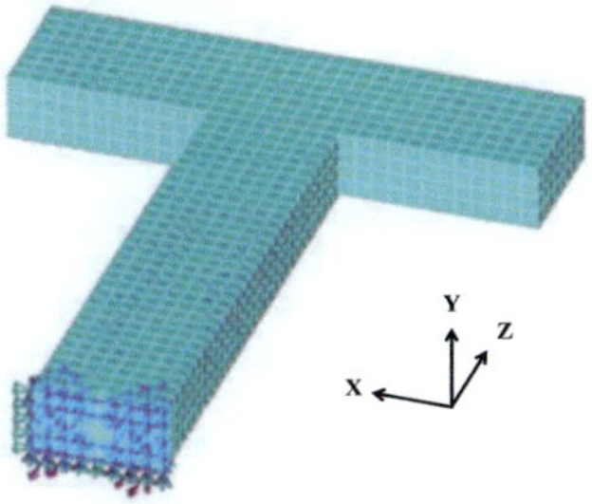

Fig. 3.2: "T" shape sample for a parameter study on the influence of the geometry on the resonant frequencies by FE simulation.

Considering a beam with a length l, a base b, and a height h, the resonant frequency is proportional to the ratio of the stiffness k over the mass m. k can be replaced by $3EI_y/l^3$, with E the material Young's modulus, I_y the second moment of area, and l, the length of the beam.

$$f = \frac{1}{2\pi}\sqrt{\frac{k}{m}} = \frac{1}{2\pi}\sqrt{\frac{3EI_y}{ml^3}} = \frac{1}{2\pi}\sqrt{\frac{3E\frac{bh^3}{12}}{ml^3}} = \frac{1}{2\pi}\sqrt{\frac{Ebh^3}{4\,ml^3}} \qquad \textit{Eq. 3.1}$$

From this equation, it is possible to conclude that the larger and thicker the beam is, the bigger the resonant frequency is. On contrary, the longer the beam is, the smaller the bending resonant frequency is.

Using the same way of thinking for torsion, the resonant frequency can be calculated. In this case, it is proportional to the ratio of the coefficient of torsion spring *k* over the polar moment of inertia along the axis of the beam *Ip*. This time *k* is proportional to $\mu I_t/l^3$ where μ is the shear modulus, I_t the torsion constant, and *l*, the sample length.

$$f = \frac{1}{2\pi}\sqrt{\frac{k}{I_p}} = c\sqrt{\frac{\mu I_t}{I_p l^3}} = c_1\sqrt{\frac{\mu b h^3}{bh(b^2+h^2)l^3}} \qquad \textit{Eq. 3.2}$$

Where c and c_1 are constants.

It is more complicated to establish a link between the sample geometry and the resonant frequency, and this is why a parameter study has been done on the length and width of the "T" shape sample.

Fig. 3.3 shows the influence of the sample length of the vertical line of the "T" on the bending and torsional resonant frequencies. With a reasonable sample length of 10 mm, the bending resonant frequency can be around 1 kHz. On the other side the torsional resonant frequency is still too high for a multiaxial mode, where the bending and torsional modes are close to each other.

The influence of the width of the horizontal line of the "T" (corresponding to the mass) should have a bigger influence on the torsion.

The width of the horizontal line of the "T" has an influence on both resonant frequencies. The influence on the torsional mode is as expected bigger than on the bending one. The frequencies cross in this example (**Fig. 3.4**) at around 30 kHz for a width of 2 mm, and converge to 2 kHz for a width longer than 10 mm.

However a "T" shape is not a realistic shape for a sample. The horizontal line of the "T", representing the mass, cannot be too long otherwise the sample handling is too difficult.

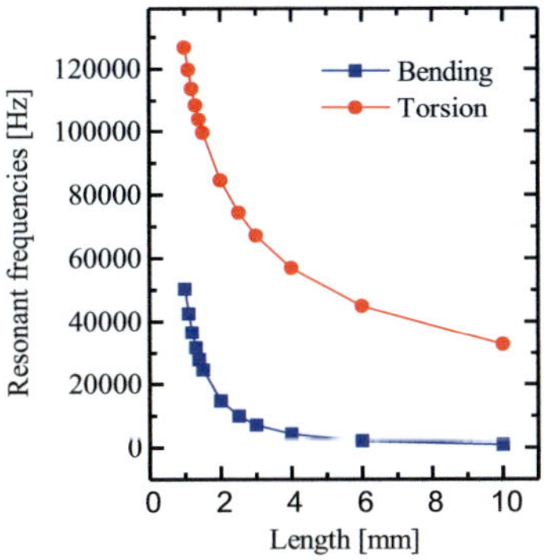

Fig. 3.3: Sample length influence on the bending and torsional resonant frequencies.

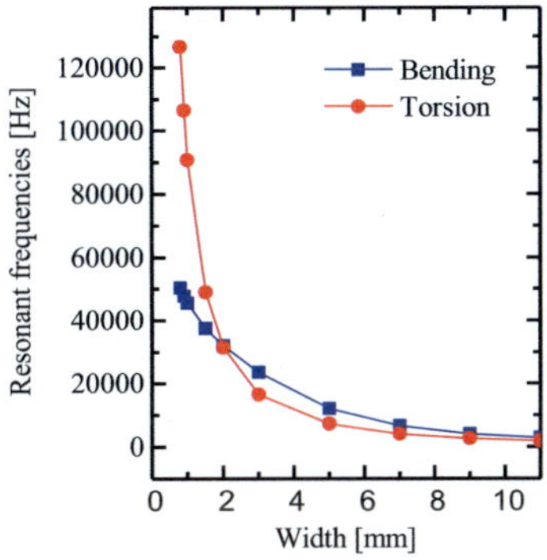

Fig. 3.4: Sample width influence on the bending and torsional resonant frequencies.

3.2.2 "T" shape and two masses sample geometry parameter pre study

Two masses (**Fig. 3.5**) at the end of the "T" shape sample allow for an increase of the mass, and by this way allow for a decrease of the resonant frequencies.

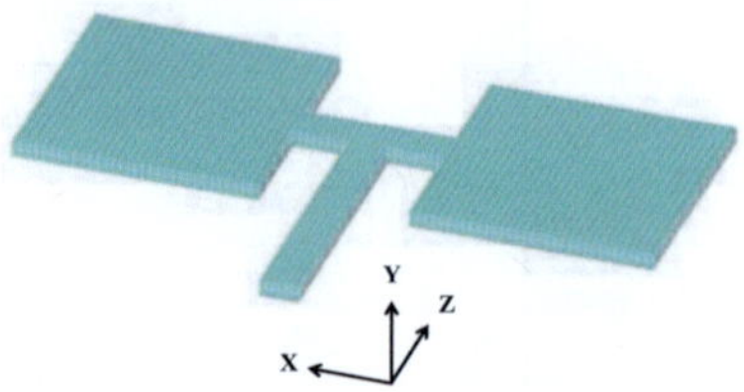

Fig. 3.5: "T" shape and two masses sample. The masses have a large impact on the resonant frequency decrease.

Based on this geometry, the two resonant frequencies intersect at around 860 Hz (**Fig. 3.6**). The basic geometry idea has been found. By adding an area where the sample can be clamped, and optimizing the sample shoulder, an ideal geometry can be generated.

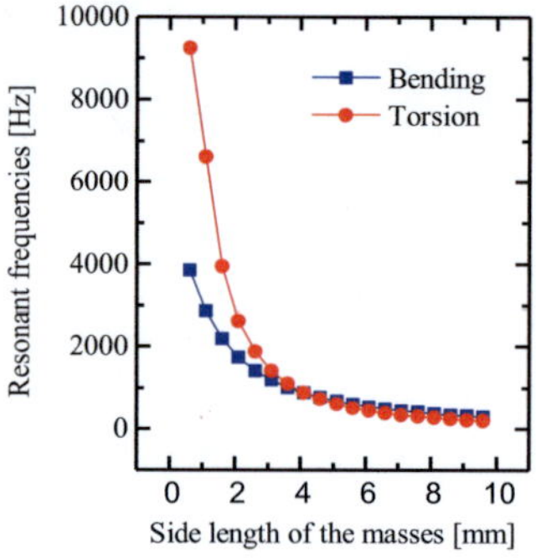

Fig. 3.6: Length influence of the sample with two masses on the bending and torsional resonant frequencies.

3.2.3 FE conditions

Two kinds of Finite Element (FE) analyses were conducted. The static analysis allowed for an estimation of the stress distribution in the sample under a certain deformation. In this case, an out-of-plane displacement *Uy* was applied to simulate bending, with an angular rotation α of the mass around the Z axis of the sample to simulate a torsional deformation. The second analysis was modal in nature, and is typically used to determine the resonant modes of the model. It can also serve as a first step for more complex analyses such as the harmonic response of the sample or a full transient dynamic analysis. Furthermore, it is a necessary input for multiaxial fatigue models. All simulations have been carried out with ANSYS using various mesh elements, and SOLID 185 was chosen as the best option. SOLID 185 is used for 3-D modeling of solid structures, and is defined by eight nodes having three degrees of freedom at each node. This element type allows for a smooth transition between volumes with different mesh sizes. With this element, the stress extraction is possible at each node and allows for having comparable node and element solutions converging to the same value. Finally, isotropic elastic properties were chosen for each sample material.

3.2.4 Sample geometry idea

When a fatigue sample is designed to work at its resonance, the sample geometry must be optimized considering restrictions from the setup. To achieve multiaxial loading conditions, the sample needs to be designed in a way that the torsional and bending modes do have a similar resonant frequency.

In the following section, the design and optimization of such samples is introduced. The samples can be divided into three sections: the gripping section, the gage section and the mass.

3.2.4.1 The gripping section

The gripping section is represented at the top of **Fig. 3.7**, it is settled into the clamps of the setup. The design requires a firm grip which does not introduce friction which would influence the measurement. It has an 11 x 6 mm rectangular shape.

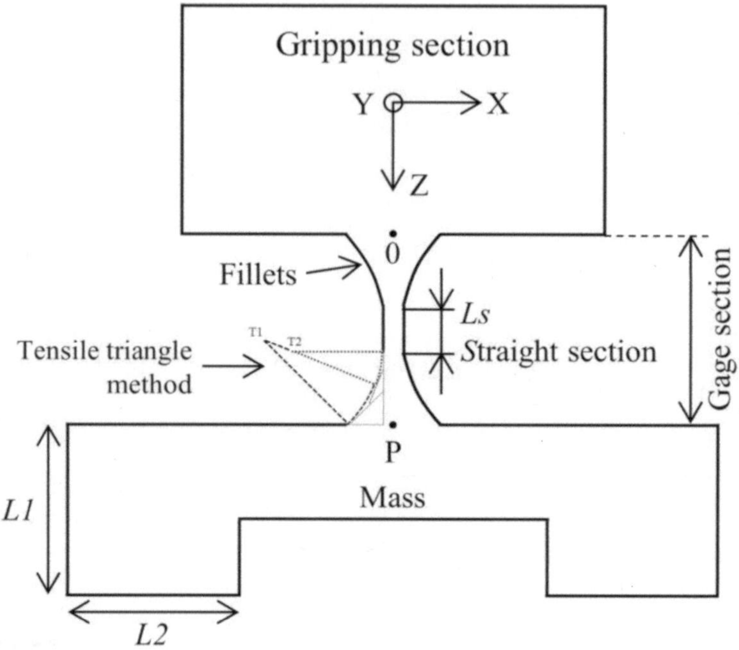

Fig. 3.7: Details of the sample geometry consisting of clamp, gage section and mass. The sample thickness is uniform.

3.2.4.2 The gage section

The gage section has a typical length of 5 mm. The shoulder fillets have been optimized using C. Mattheck's method of tensile triangles [57]. The aim is to reduce the notch stresses where the fillet joins the straight gage section. Such localized stress concentrations can lead to the initiation of cracks and failure. Due to the bio-inspired Mattheck's geometric optimization, it is possible to reduce these stresses at the notch by deriving the fillet with the usage of a least three stacked triangles, starting with the largest radius (4.89 mm) at the fixation, and ending with the smallest radius (3.49 mm) joining the gage section (**Fig. 3.1**b, **Fig. 3.1**d). The straight section with a length *Ls* is situated between the lower and the upper shoulder fillets. Two gage section geometries with the same cross section were simulated. The first has a straight section of *Ls* and the second an hourglass shape with *Ls* = 0 µm. The two simulations show the maximal bending and torsional stresses at the same point at the end of the two lowest fillets. This tensile triangle optimization has been parameterized, allowing for a modification of the geometry of the sample in the software ANSYS. Knowing the length of the sample straight part *Ls* and the total length of the sample gage (2 x *h* = 5 mm), it is possible to calculate the fillet width *x* as a function of the parameters *h* and *Ls* (**Fig. 3.8**).

In the lowest triangle, *y* can be found as a function of *x*, using the Pythagorean Theorem:

$$x^2 + x^2 = (2y)^2 \quad \rightarrow \quad y = x\,\frac{\sqrt{2}}{2} \qquad \textit{Eq. 3.3}$$

In the intermediate triangle, *z* can be found as a function of *y* and then of *x*, using the Theorem of Al-Kashi:

$$(2z)^2 = y^2 + y^2 - 2\,y\,y\cos\frac{3\pi}{4}$$

$$\rightarrow \quad 4z^2 = 2\,y^2\left(1 + \frac{\sqrt{2}}{2}\right)$$

$$\rightarrow \quad z = \frac{x}{2}\sqrt{1 + \frac{\sqrt{2}}{2}} \qquad \textit{Eq. 3.4}$$

Finally,

$$x = h - \frac{Ls}{2} - z - y$$

$$\rightarrow \quad x = h - \frac{Ls}{2} - \frac{x}{2}\sqrt{1 + \frac{\sqrt{2}}{2}} - x\,\frac{\sqrt{2}}{2}$$

$$\rightarrow \quad x = \frac{h - \frac{Ls}{2}}{1 + \frac{\sqrt{2}}{2} + \sqrt{1 + \frac{\sqrt{2}}{2}}} \qquad \textit{Eq. 3.5}$$

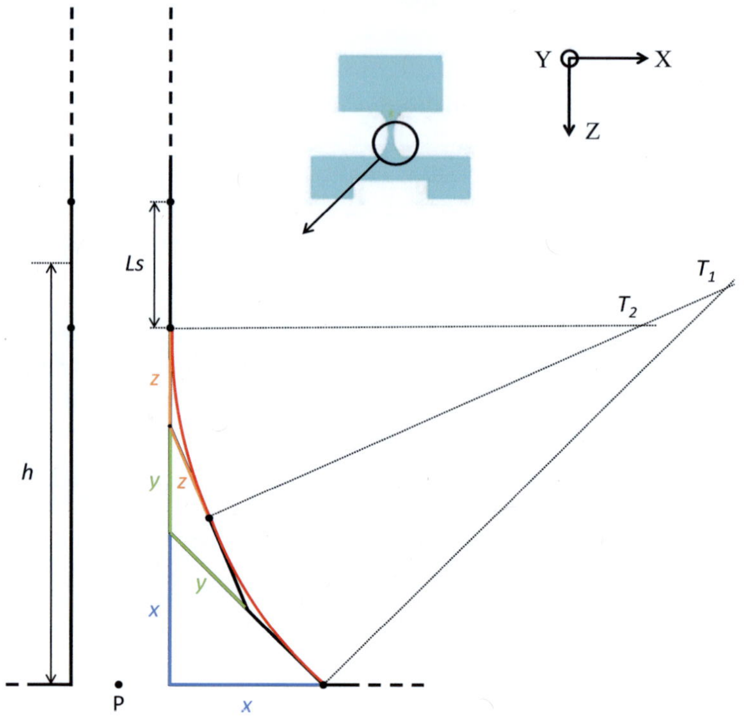

Fig. 3.8: Optimization of the shoulder fillets with the tensile triangle method to reduce the notch stresses.

3.2.4.3 The mass

Together with the stiffness of the gage section, the geometry of the mass defines the resonant frequencies of the sample. Therefore, an optimization has been conducted by the help of an FE analysis (ANSYS®, United States of America (USA)) and MATLAB mathematical software (Mathworks®, USA) to find the length *L1* and the width *L2* of the two rectangles (**Fig. 3.7**). These parameters define the two resonant modes (bending and torsion) of the sample at frequencies very close to each other. The length *L1* and the width *L2* of the two rectangles were equally varied in order to find the frequency at which the bending and torsional modes cross. Changing *L2* does not change the width of the sample, but only the width of the rectangles.

For the proof of concept, Ni-samples with a thickness of 200 μm were designed and the results are shown in **Fig. 3.9**. The point at which the bending and torsional modes cross was found to be at 460 Hz for a length of $L = L1 = L2 = 4.5$ mm.

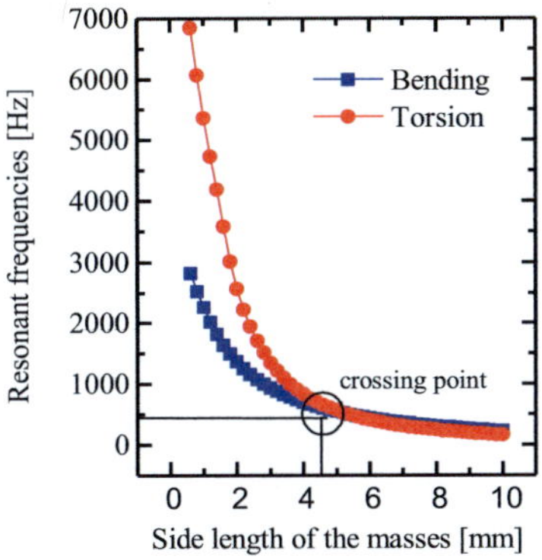

Fig. 3.9: FE optimization of the mass of a Ni-sample where $L = L1 = L2$ – resonant frequencies of bending and torsional modes are equal at the crossing point.

The disadvantage to this design is that the resonant frequency is rather low (460 Hz), and more than 25 days would be necessary to reach 10^9 cycles. Therefore, consecutive optimizations were performed with the goal of reaching the kHz regime. The geometry of the mass plays an important role on the resulting resonant frequencies of the sample. With an increase in the length *L1* of the mass, a decrease in the bending mode resonant frequency can be expected. *L1* also has an impact on the torsional resonant frequency, but it is significantly lower. In the same way, the larger the width *L2* is, the lower is the torsional frequency, while a minor change on bending frequency results. Nevertheless, both modes are influenced by the total mass and the heavier the mass, the lower is the resonant frequency.

Based on this behavior, an optimization has been conducted by the help of an FE modal analysis (ANSYS) to find the length *L1* and the width *L2* of the two rectangles which leads to achieve high frequencies for the resonant modes (bending and torsion) with close proximity. In case of $L1 = L2 = L$, the simulation shows that the torsional mode frequency decreases faster than the bending mode frequency when L increases. Therefore *L1* and *L2* are defined such as: $\begin{cases} L1 = x \times L \\ L2 = L \end{cases}$ with $x > 1$. For each value of x, there is one value of the length L where the torsional mode frequency and the bending mode frequency are nearly equal. By plotting the multiaxial mode frequency over L, the mass geometry can be found for a designed resonant frequency (**Fig. 3.10**).

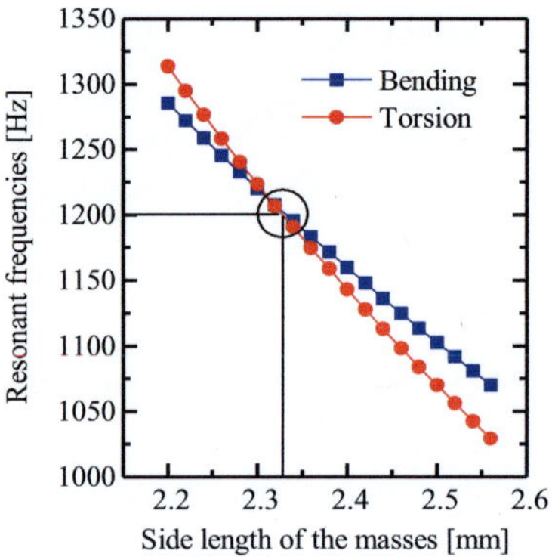

Fig. 3.10: FE optimization of the mass of a Al-sample where $L = L1 \neq L2$ – resonant frequencies of bending and torsional modes are equal at the crossing point.

Another design was optimized for Al and Cu to reach a resonant frequency in the kHz regime for VHCF testing. The sample geometry has the following dimensions:

in the case of the Al-samples: $$\begin{cases} L1 = 2 \times L = 4.64\ \text{mm} \\ L2 = L = 2.32\ \text{mm} \\ f_{\text{multiaxial}} = 1207\ \text{Hz} \end{cases},$$

in the case of Cu-samples: $$\begin{cases} L1 = 8 \times L = 6.32\ \text{mm} \\ L2 = L = 0.79\ \text{mm} \\ f_{\text{multiaxial}} = 1180\ \text{Hz} \end{cases}.$$

This method is not the most efficient. It takes time and demands many calculations. However, it results in a geometry of the mass that respects the requirements: a high resonant frequency. A potential future study would be to write a MATLAB routine which calculates through ANSYS an appropriate geometry from given material parameters.

Fig. 3.11 shows an FE frequency sweep. The displacement amplitude of the sample at position P (**Fig. 3.7**) is plotted versus frequency. The first peak at 454 Hz corresponds to the torsional mode, the second one at 463 Hz to the bending mode. In between, at 460 Hz, the sample can be excited in a 'multiaxial mode'. Therefore, three different kinds of resonant fatigue tests can be conducted with this design: pure bending, pure torsion and a combined loading with an adjustable phase angle.

Another simulation has been done to evaluate the influence of the clamping of the sample. **Fig. 3.12** shows the new applied boundary conditions with 1 mm unconstrained on top of the gripping section. Compared to the previous analysis, these changes on the boundary conditions affect the sample resonant modes by less than 0.2%.

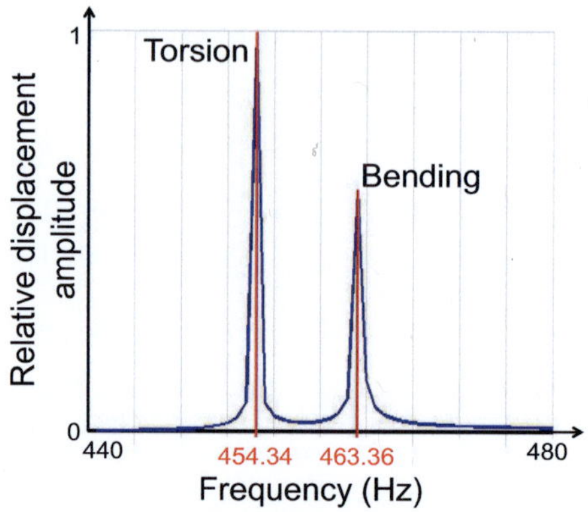

Fig. 3.11: FE simulated displacement amplitude of a Ni-sample for a frequency sweep showing two peaks, one for bending and one for torsion.

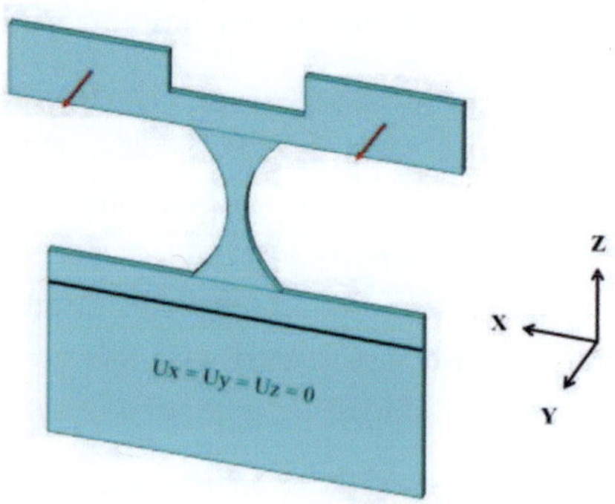

Fig. 3.12: Sample model for studying the influence of the clamping conditions.

3.2.5 Final sample geometry

The geometry optimization has been done for several materials, thicknesses, frequencies (around 500 Hz for Ni, or 1 kHz for Al and Cu) and geometries (with an hourglass shape or not). The following table (**Tab. 3.1**) summarizes the data of the four final samples.

Sample geometry	**Material thickness**	**Multiaxial mode frequency (simulated)**	**Geometry parameters**
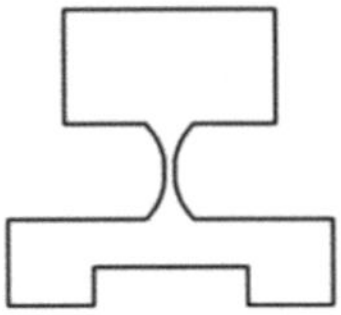	Ni 200 μm	460 Hz	Ls = 0.3867 mm $T1$ = 8 mm $T2$ = 2.5 mm $L1$ = 4.5 mm $L2$ = 4.5 mm
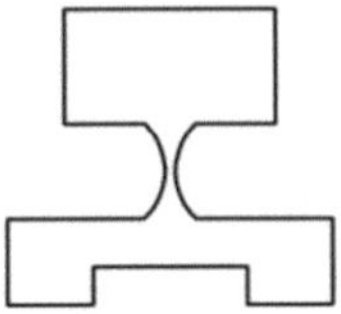	Al 250 μm	1207 Hz	Ls = 0 mm $T1$ = 4 mm $T2$ = 1 mm $L1$ = 4.64 mm $L2$ = 2.32 mm
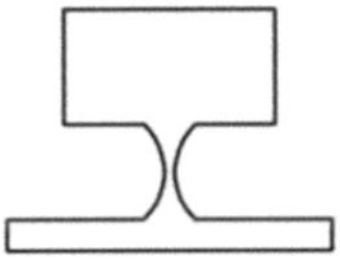	Cu 200 μm	1180 Hz	Ls = 0mm $T1$ = 0 mm $T2$ = 0.79 mm $L1$ = 6.32 mm $L2$ = 0.79 mm
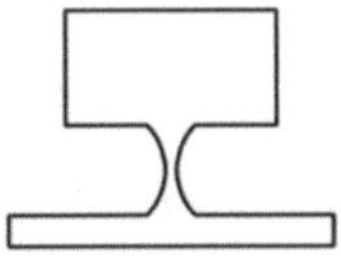	Cu 100 μm	1147 Hz	Ls = 0 mm $T1$ = 0 mm $T2$ = 0.53 mm $L1$ = 5.3 mm $L2$ = 0.53 mm

Tab. 3.1: Final sample geometries depending on the material. Gage section length equal to 5 mm. The sample thickness is uniform.

In conclusion, the modal analysis allowed for determining the sample geometries for which the bending and torsional resonant frequencies are around 1.2 kHz and close enough to each other that a multiaxial mode can be excited. Moreover, the hourglass shape is more relevant for a multiaxial mode considering the fact that the areas of maximal stress for bending and torsion are closer than for the samples with a straight gage section.

3.2.6 Boundary conditions

In the simulations, all degrees of freedom in displacement were set to zero at the two gripping surfaces. In the static analysis, bending was simulated with an out-of-plane displacement in the Y direction applied on the nodes at the inner edge of the mass (**Fig. 3.13**a). Torsion was simulated with an angular rotation α of the mass around the Z axis of the sample. Therefore, all nodes on the same line were displaced of Uy, depending linearly on X, and being zero at point P. In this case, the displacement depends on the coordinate Xn of each node n: $Uy = \tan(\alpha) \times Xn$ (**Fig. 3.13**b). In the modal analysis, the gripping section is also fixed but no further boundary conditions are applied.

To reduce computational cost, the sample has been cut into three parts: the mass and the gripping section with the same coarse element size, and the double-filleted gage section with a smaller element size. The element sizes were chosen based on a compromise between the convergence of the results and a reasonable computation time. In the gage section, two elements were used through the sample thickness. For the gripping and mass sections, a single hexahedral element was sufficient to model the sample thickness. Using these element sizes and meshing the sample with an automatic sweep mesh, the resonant frequency value converges with a 0.1% error.

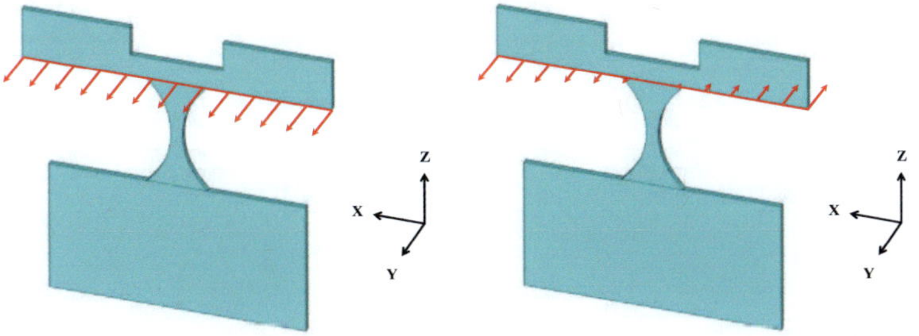

(a) (b)

Fig. 3.13: Displacement applied in the static FE-simulation of the sample for a bending (a) and for a torsional (b) loading.

3.2.7 Stress state correlation between static and modal analysis

The displacement of the sample exposed to its resonant frequency determines the stress state. It is now important to validate the stress/displacement distributions obtained by ANSYS from a static or a modal analysis in order to ascertain the stress in the sample.

Indeed, in comparison with a static analysis, a modal analysis has the most realistic deformed shape and corresponds to the behavior of the sample in test. Nevertheless, in ANSYS, a coherent damping of the system is not taken into account, and because of this, stresses and displacements for each node or element are completely out of scale (e.g. displacement of 148 m with a modal analysis, **Fig. 3.14**c). However these results still make sense, and have to be used as a ratio of the stress over the displacement.

Finally, the comparison of a static and a modal analysis show that the stress distributions are identical (**Fig. 3.14**).

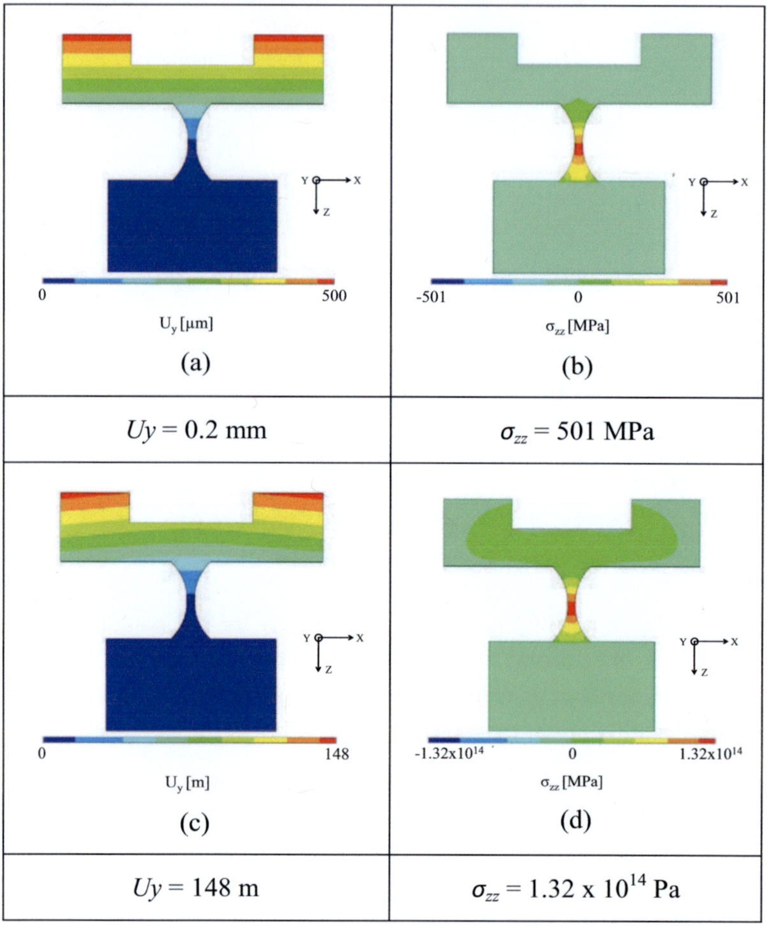

Fig. 3.14: Out-of-plane displacement U_y obtained from a static analysis (a) and a modal analysis (c), and the corresponding stress distribution (b) and (d), respectively.

A displacement of $Uy = 200$ µm leads to a maximal stress of $\sigma_{zz} = 501$ MPa in the static analysis, while a displacement of $Uy = 200$ µm leads to a maximal stress of $\sigma_{zz} = 512$ MPa by scaling the results from the modal analysis. The relative difference between both results is 2%. It is considered as a small error and validates the stress/displacement ratio from the modal analysis.

3.3 Implementation of the resonant fatigue setup

3.3.1 Setup principle and hardware

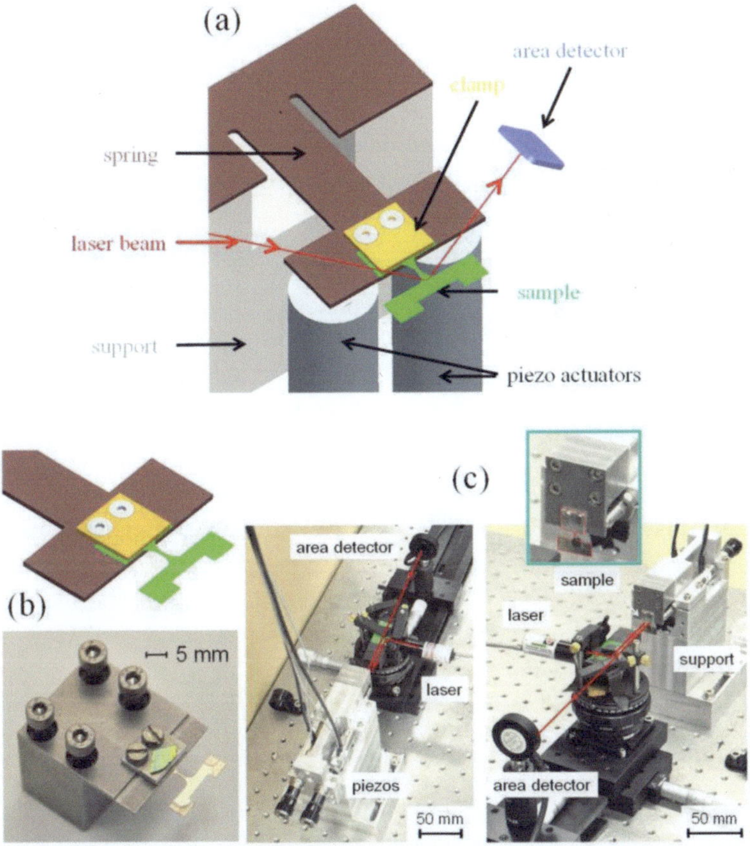

Fig. 3.15: (a) Computer-aided design (CAD) of the multiaxial setup, (b) details of sample and clamp, (c) picture of the setup.

A novel custom-built fatigue setup was developed to investigate the fatigue behavior of micro-samples under cyclic multiaxial loading in the VHCF regime. The design of the resonant fatigue setup (**Fig. 3.15**a) consists of two piezo actuators whereupon a flat spring is clamped. The actuation is implemented by two piezo stacks (P-843.40, Physik Instrumente, Germany), each one with a travel range of 60 μm, and a resolution of 1.2 nm in closed loop. These piezo stacks are high resolution linear actuators intended for static and dynamic applications. They provide a sub-millisecond response time, which is necessary for kHz actuation. Each piezo actuator is equipped with integrated high-resolution position sensors (linearity to 0.01%, and a bandwidth up to 5 kHz). These sensors allow for measuring the two piezo displacement signals, i.e. the exact setup actuation.

The spring can be slightly bent and/or twisted by the two actuators. It is made out of steel and its geometry was optimized with FE, and chosen in a way that the surface strain is always well below the fatigue limit while its resonant frequency is well above the sample resonant frequency. With this optimization, fatigue phenomena never occur in the spring.

In the same way as for the sample geometry in 3.2.1, an optimization has been done for the spring where each parameter (length, base, height) has been changed. Since the spring is part of the clamping system, its bending resonant frequency has to be higher than the one from the sample (around 1 kHz). A higher resonant frequency is desirable to avoid fatiguing the clamp rather than the sample. Two models (**Fig. 3.16**) corresponding to the requested properties have been found.

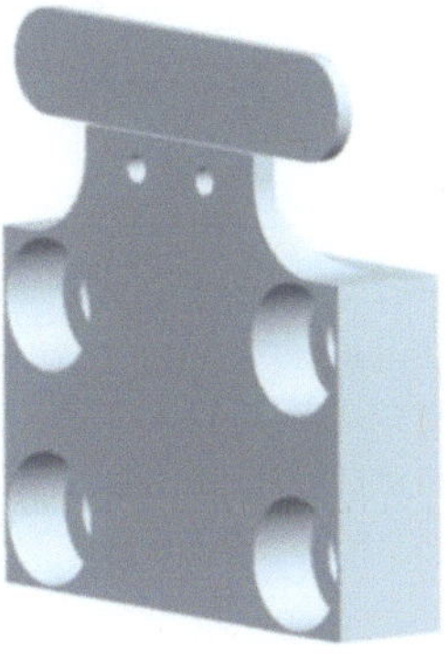

(a) 1100 Hz

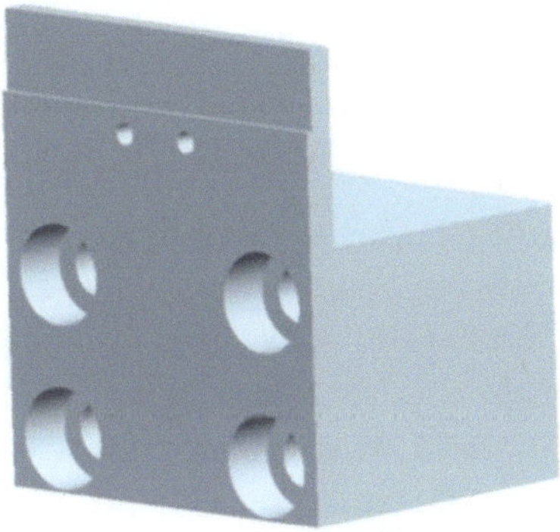

(b) 8900 Hz

Fig. 3.16: CAD models of the clamping system, and their first bending resonant frequency.

Moreover, the piezo actuators have to be able to bend the clamping system. The compression force has to be less than 800 N for a P-843.40 (Physik Instrumente, Karlsruhe, Germany). The maximal amplitude of the clamp is equal to the maximal displacement of the piezo actuators. The required load *F* at the unsupported end can be calculated this way:

$$F = \frac{3dEI}{l^3} \qquad \textit{Eq. 3.6}$$

where d is the deflection, E the Young's modulus of the clamp material, I the second moment of area, and l the length of the clamp.

In this case, according to the second clamp geometry (**Fig. 3.16**b) and the piezo actuators properties: $d = 60$ µm, the maximal amplitude of the piezo actuators; $E = 210$ GPa, the Young's modulus of steel; $I = \frac{bh^3}{12}$, with $b = 30$ mm, $h = 1$ mm and $l = 15$ mm, F is equal to: $F = 28$ N.

F is under the 800 N limitation, there are no risks for the piezo actuators to be damaged by the clamp. The geometry in **Fig. 3.16**b of the clamp has been chosen to be used in the setup since its simple geometry allows for a fast and easy machining.

Fig. 3.17 shows the different possible positions of the clamping system and of the sample in the setup. First, the sample can be oriented upward (a); second, the clamping system is arranged such that the sample is excited with only one piezo actuator (b). Since the gravity could have an effect on the result, the frame has been modified in order to fix the sample downward (c).

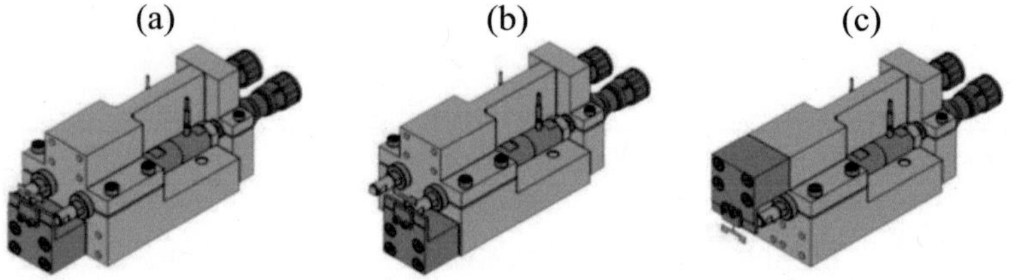

Fig. 3.17: CAD of the multiaxial setup: (a) upward sample position, (b) sample excited with only one piezo actuator, (c) downward sample position.

The displacement amplitude is measured by the use of a laser beam. The beam is reflected from the surface of the sample at the point P (**Fig. 3.7**) onto an area detector. The area detector is a Position Sensitive Device (PSD), and has a high frequency analog data acquisition, an update rate up to 30 kHz, and an optical resolution of 1 μm. This resolution corresponds to a stress amplitude measurement error of 0.1% when considering a nickel sample in the described setup. The sensor measures the bending and torsional deflection of the sample; i.e. the exact output oscillation.

The experimental control and data acquisition software was programmed in LabVIEW (National Instruments, USA). It mainly runs on a Field Programmable Gate Array (FPGA) (NI PCIe-7852R, National Instrument, USA) which can digitize and process data from eight analog inputs at a sample rate of up to 750 kHz (analog input 750 kSample/s and analog output 1 MSample/s, clock rate 40 MHz). Hence, the experiments can be stopped automatically in ms when the fatigue criterion, e.g. a frequency decrease, is exceeded. Thus, high speed load control and data acquisition are both capabilities of the setup. As an example, the two piezo actuators are controlled by the FPGA following this way: $AO_0 = A_1 \sin(2\pi f t) + (-A_2)\sin(2\pi f t - \theta)$ and $AO_1 = A_1 \sin(2\pi f t) + A_2 \sin(2\pi f t - \theta)$. With AO_i being the analog outputs from the FPGA, which are the control voltages for the piezo actuators; A_1 is the bending amplification parameter; A_2 is the torsion amplification parameter, f the driving frequency, t the time, and θ the phase angle between bending and torsional actuations. If $A_2 = 0$, AO_0 and AO_1 are equal, the two piezo actuators are working in-phase. A_1 has a high influence on the bending mode of the sample. On the other hand, if $A_1 = 0$, AO_0 and AO_1 are additive inverse, the two piezo actuators are out-of-phase. The sample can reach its torsional mode. If $A_1 \neq 0$ and $A_2 \neq 0$, the combination of bending and torsion results in a so called ‘multiaxial mode’.

Every 500 ms, the displacement data is analyzed (tone measurement) for the amplitude of the dominating frequencies in bending and torsion, as well as the phase angles between the two modes and the two actuators. From this analysis the sinusoidal excited displacement amplitude are adjusted by a Proportional-Integral-Derivative (PID) control. The parameters P, I, and D were found through the application of the Ziegler-Nichols and Tyreus-Luyben tuning methods. The necessary procedure to find the right PID parameters is described in the chapter ‘Experimental Implementation’.

The amplitude PID and frequency controllers implemented in LabVIEW enable amplitude and frequency control of the piezo driven excitations that occur in the sample, allowing for multiaxial fatigue at resonance. The control process addresses two key phase angles: θ between the two actuation modes, and Φ between the piezo actuation and sample oscillation. Controlling θ allows for changing the phase shift between the torsional and bending mode. The sample phase shift,

Φ, is calculated from an actuation frequency and a displacement measurement made by the area detector. This measurement is then used as the primary plant input for the amplitude and frequency controllers. PID control regulates piezo amplitudes and thus the stress amplitude of the sample, while frequency control is responsible for keeping Φ fixed at 90° corresponding to the resonance. Restated, the sample response measured by the surface detector acts as the primary input for the amplitude PID and frequency controllers; i.e., as measured displacements in the sample deviate from the desired amplitude, the controllers regulate the piezo actuation until the resonant frequency and target stress state are attained.

3.3.2 Stress calculation for the sample

With the custom-built resonant fatigue setup used for the experiments, different modes of loading can be chosen by changing the driving frequency and amplitudes. **Fig. 3.18** represents the sample cyclic displacement (in V) on the area detector during experimentation. It is possible to distinguish three different modes. The first one corresponds to the bending mode, shown in blue. A perfect bending displacement would be a totally flat ellipse coinciding with the ordinate axis. In the same way, a perfect torsion displacement would be a flat ellipse coinciding with the abscissal axis (red in **Fig. 3.18)**. All other possible ellipses represent a multiaxial displacement. The line in purple corresponds to an in-phase multiaxial displacement with exactly half torsion and half bending (angle of 45°).

The black circle (dashed line) represents the out-of-phase mode, where bending is at its maximum/minimum when torsion is at zero, and by the same way at zero when torsion is at its maximum/minimum. Finally, the purple dashed line shows an example of a multiaxial state, with a certain amount of bending and torsion.

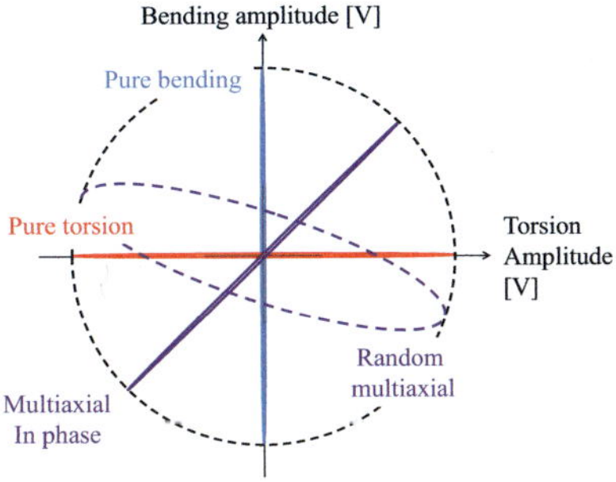

Fig. 3.18: Different possible cyclic displacement of the reflected laser beam on the area detector.

The stress amplitude of the sample gage, under a bending or a torsional loading, is calculated from this measured displacement amplitude of the sample at its free end at position P (**Fig. 3.7**). The displacement is compared to a structural static finite element analysis in order to determine the stress in the middle of the sample (**Fig. 3.1**b, **Fig. 3.1**d).

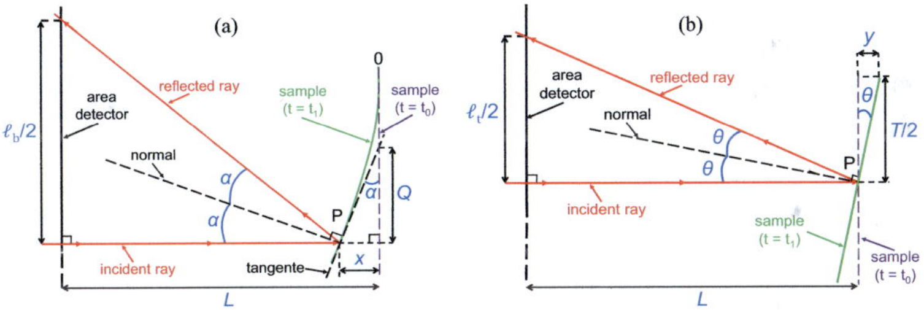

Fig. 3.19: Geometric calculation of the sample displacement x for bending (a), and of the angle of rotation θ for torsion (b).

The details of the geometric calculation for the bending mode are illustrated in **Fig. 3.19**a. The laser is pointed onto the sample at position P. Q is the length between "the intersection of the tangent on P of the deformed sample and the non-deformed sample" and "the point P from the non-deformed sample". The static and modal analyses in ANSYS show that the parameter Q remained constant for a typical amplitude range applied during testing. The angle between the non-deformed sample and the tangent on point P of the deformed sample is called α. The laser beam is reflected with an angle of 2α. Finally, $l_b/2$ is the displacement amplitude of the laser beam measured on the area detector. The same method is used for torsion. θ is the sample torsional angle (**Fig. 3.19**b), T the sample width and $l_t/2$ the laser beam displacement measured on the area detector.

Using trigonometry and the bending and torsional deflections measured by the area detector, the real displacements x (**Fig. 3.19**a) and y (**Fig. 3.19**b), corresponding respectively to α and θ of the sample can be calculated. Due to the bending deformation, the measured displacement varies depending on the position of the laser point on the sample. In case of a large bending amplitude, the laser beam intersects with the sample at different positions depending on the deflection. However, for the typical bending amplitudes used in this thesis this corresponds to an error of 0.2% which can be neglected.

In **Fig. 3.19**: the angle α can be calculated in the triangle formed by the tangent between the displacement x at position P and the length Q:

$$tan(\alpha) = \frac{x}{Q} \qquad \textit{Eq. 3.7}$$

This angle corresponds to the deflection of the laser and the point of intersection on the detector:

$$tan(2\alpha) = \frac{\ell_b}{2(L-x)} \qquad \textit{Eq. 3.8}$$

Taking into account that

$$tan(2\alpha) = \frac{2\,tan(\alpha)}{1-tan^2(\alpha)} \qquad \textit{Eq. 3.9}$$

it can be deduced that

$$\frac{\ell_b}{2(L-x)} = \frac{\frac{2}{Q}x}{1-\frac{1}{Q^2}x^2} \qquad \textit{Eq. 3.10}$$

and

$$(\ell_b - 4Q)x^2 + 4LQx - \ell_b Q^2 = 0 \qquad \textit{Eq. 3.11}$$

The discriminant of this polynomial is:

$$\Delta = 4Q^2({\ell_b}^2 - 4\ell_b Q + 4L^2) \qquad \textit{Eq. 3.12}$$

where one solution is positive:

$$x = \frac{Q\left(\sqrt{{\ell_b}^2 - 4\ell_b Q + 4L^2} \quad 2L\right)}{\ell_b - 4Q} \qquad \textit{Eq. 3.13}$$

The rotational motion of the sample mass, resulting from the torsional deformation of the sample gage section, can be measured by the laser interception with the detector y. For pure torsion, the angle θ can be calculated on the sample side (**Fig. 3.19**b):

$$\tan(\theta) = \frac{y}{T/2} \qquad \textit{Eq. 3.14}$$

and on the detector side by:

$$\tan(2\theta) = \frac{\ell_t}{2L} \qquad \textit{Eq. 3.15}$$

It can be deduced that:

$$y = \frac{T}{2}\tan(\theta) \qquad \textit{Eq. 3.16}$$

with

$$\theta = \frac{1}{2}\arctan\left(\frac{\ell_t}{2L}\right) \qquad \textit{Eq. 3.17}$$

The deformed shape of the geometrically complex sample is analyzed by an FE modal analysis through which the variable Q is calculated. A MATLAB routine is used to launch ANSYS and calculate the maximal stress in the center of the gage section depending on the displacement of the deflected laser beam ℓ_t and ℓ_b.

Finally, the different systematic errors and uncertainties in this setup should be evaluated. The main errors stem from the determination of the sample geometry, the surface quality, and the position of the sample, as well as the measurement error of the sensors. For example, the measurement error for the sample thickness correlates linearly with the error of the calculated stress amplitude. The thickness is determined with a digital optical microscope and has a resolution of 1 pixel, corresponding to 0.18 µm (0.09% of a 200 µm thickness). Furthermore, the surface of the mass section needs to be absolutely planar so that the laser beam is reflected as described in the sections above. It is difficult to estimate this error, but experience suggests that samples with a plane surface at the mass are not showing measurable deviations. The distance between the sample and the area detector L is also linearly positioned. This distance is controlled by a micro screw with a resolution of 10 µm and corresponds to a 0.06% error. Another error is introduced by the area detector, which can detect the reflected beam position with a resolution of 1 µm, corresponding to a 0.03% for a typical VHCF test. Also, if the surface roughness of the sample is too high, the reflected beam is broadened and parts of the reflected beam could lie outside the sensor area. This error can be prevented by limiting the sensor range and optimizing the optical path so that the reflected laser beam stays on the area detector. The displacement sensor in the piezo actuator has a nominal resolution of 1.2 nm for their 60 µm displacement range (0.002% error). This is the smallest displacement amplitude size the PID is able to control.

To prevent timing issues, all signals are routed as analog signals and are digitized at the same time by the FPGA using the same reference clock. The FPGA clock is working at a frequency of 40 MHz. The time resolution is therefore determined by the FPGA and is in the range of 25 ns. In theory, there could be time lags for the different length of cables, but relative to the time resolution, they are negligible.

3.3.3 PID control of the displacement amplitude

Since the sample is driven at resonance, the control of the amplitude becomes a critical issue when fatigue damage occurs. Therefore, a precise control of the amplitude is of absolute necessity for the quality and reliability of the experiments. Due to the system low response time at resonance, the control is performed only every 500 ms by a PID controller, a type of closed-loop feedback regulation. For this particular fatigue setup, two output signals are regulated: the bending and torsional amplitudes. The standard governing equation for PID controllers is given by [58]:

$$u(t) = K_p \left[e(t) + \frac{1}{T_i} \int_0^t e(\tau) d\tau + T_d \frac{d}{dt} e(t) \right] \qquad \textit{Eq. 3.18}$$

Where $u(t)$ is the modified variable output, $e(t)$ is the error between desired signal and measured signal, K_p is the proportional gain, T_i is the integral time, and T_d is the derivative time. A systematic approach to finding these PID constants was desired to increase process reliability. A PID control with quarter wave decay was achieved through the application of the Ziegler-Nichols method, and the Tyreus-Luyben tuning coefficients were used to enhance robustness and decrease output oscillation. The Ziegler-Nichols (ZN) method requires the user to find two values, namely, the critical gain and oscillatory period at this critical gain. To clarify, the critical gain is the value of the proportional coefficient at which the output neither converges nor diverges; rather, it remains in oscillation.

At this critical gain, one must measure the period of oscillation in order to apply the Ziegler-Nichols heuristic values. With these two values being available, one can also apply the Tyreus-Luyben (TL) state points. While the ZN and TL methods provide an excellent starting point for understanding the effects of a PID controller on the custom-built system, there is no 'one size fits all' blanket method of ensuring a well-controlled process. Neither the ZN nor the TL values resulted in stable experimental conditions. For this reason, it was necessary to further examine the results that these methods produced, and attempt to reconcile them with fatigue experimentation in a non-perfect environment. To this end, one must understand how each constant, P, I, or D, affects the process being controlled. An explanation of the behavioral characteristics of P, I, and D is required for further discussion.

For processes with frequent large amplitude disturbances, literature suggests that the proportional P gain should dominate the PID controller's response output [58]. Using a higher integral I gain results in a system that 'remembers' past errors, and is suitable in systems with long-lasting, low-amplitude perturbations. The D gain utilizes the slope of the present error to estimate future behavior of the system, and is useful in slowing very sudden and sharp changes in system output. Because the designed system in this situation is a very high cycle fatigue setup potentially working at its resonant frequency, an assumption is made that there will be low amplitude, long-duration perturbations, and perhaps every once in a while a sharp error stemming from some critical microstructural change. With this intuitive approach, a controller with plant corrections dominated by the I and D gains is a logical choice. The P gain should not be emphasized in this case, as the typical perturbations will not be uniform across the entire time and will not be high in amplitude or frequent in occurrence.

Tab. 3.2 shows the variation between the ZN, TL, and chosen constants for the fatigue setup that were found by empirical results and the above-stated logic. The above written standard form of PID equations was used in this discussion; however, the numerical implementation in LabVIEW is different due to its discretization method.

	P (K_p)	**I (T_i, min)**	**D (T_d, min)**
Ziegler-Nichols	0.307	0.0595	0.0149
Tyreus-Luyben	0.237	0.269	0.0189
Derived empirical values	0.0270	0.396	0.0287

Tab. 3.2: PID values from their respective tuning method.

It should be noted that the most important difference between the proposed values is their reliance on proportional changes. A valid fatigue test simply cannot tolerate the large changes in the stress amplitude that would occur from a proportionally dominated control system. In order to decrease the varying effects of user error on each different experimental run, the empirically discovered PID gains could be programmed into LabVIEW along with the heuristic base of the ZN and TL methods. This would allow for the user of the test setup to consistently find proper P, I, and D values for a wide range of samples (Al, Cu, or Ni with varying masses added) without the risk to alter previously consistent test conditions.

Since the setup is intended for fatigue tests at resonance, it is clear that the stress amplitude must be tightly controlled to ensure test validity. What is not so directly obvious is that the input oscillation frequency must also be controlled because fatigue damage changes the stiffness of the sample over time. Stiffness is correlated to resonant frequency, and this is tracked in the setup by the phase angle between the actuator input and the sample response oscillation.

A heavily customized proportional differential controller with adjustable slope and accuracy settings was programmed into LabVIEW to control the resonant frequency. The slope of the frequency controller can be adjusted by the user to incorporate actual sweep data in the control regime.

When the phase is controlled close to 90°, it means that the sample is being actuated at either its torsional or bending resonant frequency, and as stated, deviation from one of these modes at resonance can drastically affect the stress amplitude. Because of this behavior, the measured output frequency must be controlled to maintain resonance. While normal usage of the frequency controller is intended for a phase of 90° exactly at one of the modes, it is also often desirable to control out of mode, when strong multiaxial states are present in the sample. To give the user the option to move away from perfect modal control, a variable target phase was programmed as a function of the setup.

In this setup, the testing methodology is optimized by using the aforementioned two controllers. The resonant frequency is first defined by sweeping the driven frequency. When a rough value is found, the proportional controller for the resonant frequency is started. The driven frequency is stabilized at the sample resonant frequency by controlling the phase angle at $\Phi = 90°$. The amplitude of the sample is then slowly increased to reach the desired testing amplitude. It is important that the amplitude gradually rises. Since the resonant frequency and the amplitude are strongly correlated, an amplitude increase results in a frequency shift, which the resonant frequency proportional controller must be able to compensate for. When the testing amplitude is reached, the amplitude PID controller is started. Both controllers then work in parallel to ensure a constant resonant state and a constant amplitude. After this point, the test is in its optimized conditions and damage formation can be tracked.

It is known that the frequency, f, is proportional to the square root of the stiffness, k, over the mass, m. A reasonable assumption is made that the mass stays constant during every test. The next assumption is, that the stiffness of the sample decreases with damage formation. With the setup at hand it is possible to measure a small decrease in the resonant frequency, and thus indirectly measure the appearance of defects. A frequency sweep is done before each test (**Fig. 3.20**). Taking as an example a bending test of a Ni-sample, a change of $\Delta\Phi = 1°$ of the phase angle (between the piezo actuation and the sample oscillation) around the resonant frequency corresponds to a relative frequency change of $\Delta f / f_{\text{initial}} = 5 \times 10^{-5}$. The measured phase angle showed a noise floor of roughly $\Delta\Phi = 0.01°$. Thus the resolution of the relative change in resonant frequency is about $\Delta f / f_{\text{initial}} = 5 \times 10^{-7}$.

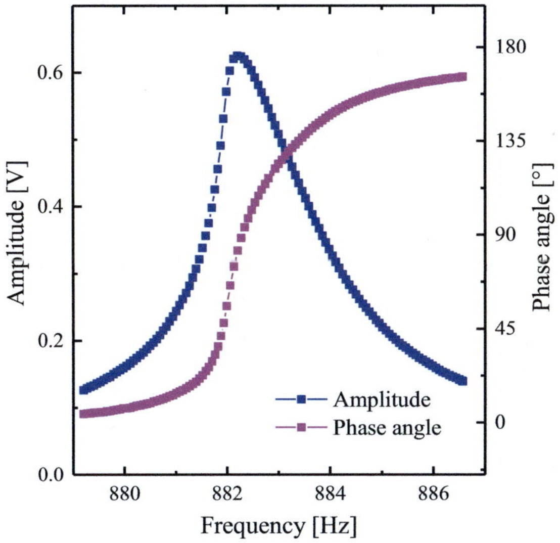

Fig. 3.20: Experimental frequency sweep for an Al-sample under a bending load.

3.4 FE implementation of a crack in a sample

To study the influence of damage on the sample resonant frequency, a crack has been implemented in the FE model. The following sections explain how the element size and crack geometry have been chosen. Different models have been developed, and used in a modal analysis to investigate the influence of cracks on the sample surface on the resonant frequency.

This study has been made over the years with the collaboration of Prof. Yuri Lapusta from the IFMA (French Institute of Advanced Mechanics) in Clermont-Ferrand. Bérengère Pierrat and Pierre Mousty, IFMA students, worked during their internship at the KIT on this FE investigation.

3.4.1 Mesh optimization

In order to obtain accurate results concerning the resonant frequencies and avoid singularities at the tips of a crack, the element size in the sample high stress region, where the crack is implemented, has to be as small as possible. Unfortunately, decreasing the element size requires progressively longer calculation time or a more powerful processor. Therefore, the sample has been cut into three parts so that the smallest element size is only used in a restraint volume. The mass and the clamp have the same coarse element size (*Esize1*) and are meshed with a structured mesh (called 'sweep mesh' in ANSYS). The two sample fillets have a smaller element size (*Esize2*) and the same structured meshes. Finally the region where the stress distribution has its maximum, typically close to the center of the sample, has the finest element size (*Esize3*), and has an unstructured mesh (called 'free mesh' in ANSYS). It allows for having elements in the µm range at the crack tip.

Finally, to avoid material, real constant, and geometric discontinuities due to the crack, the average nodal solution with the full graphics settings is selected in the ANSYS post-processing parameters.

A first MATLAB routine, starting ANSYS in batch mode, on an aluminum sample, was done on *Esize1*, with *Esize2* and *Esize3* equal to 125 µm, which corresponds to two elements along the thickness.

Four iterations (see **Fig. 3.21**) show the influence of *Esize1* on the bending and torsional resonant frequencies. The results are converging for an element size *Esize1* smaller than 130 µm for the bending frequency and 80 µm for the torsional frequency. The thickness of the aluminum sample being 250 µm, this corresponds to three elements of about 80 µm through the thickness. Therefore, *Esize1* is fixed to the sample thickness divided by three for the clamp and the mass.

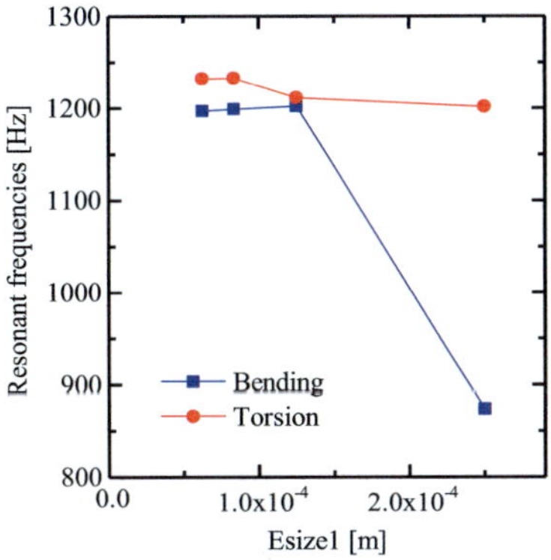

Fig. 3.21: MATLAB routine on *Esize1*, the bending and torsional resonant frequencies converge for element sizes smaller than *Esize1* = 130 µm.

A second MATLAB/ANSYS routine has been used for determining *Esize2*, with *Esize1* fixed to a third of the thickness as found in the previous paragraph, and *Esize3* fixed to 125 µm, corresponding to two elements through the thickness. The resonant frequencies are converging when *Esize2* is smaller than 35 µm (**Fig. 3.22**).

It corresponds approximately to the sample thickness divided by seven. This value is kept for the element size along the fillets (*Esize2*).

Finally, *Esize3* has been found with a third MATLAB/ANSYS routine, taking into account the previously chosen values of *Esize1* and *Esize2*. The influence of *Esize3* (**Fig. 3.23**) on the resonant frequencies is small compared to the one from *Esize1* and *Esize2*. A value for *Esize3* of 7.5 µm (sample thickness divided by 33) is considered as small enough for the convergence of the results. This corresponds to a 0.3% error if *Esize3* was equal to 1 µm.

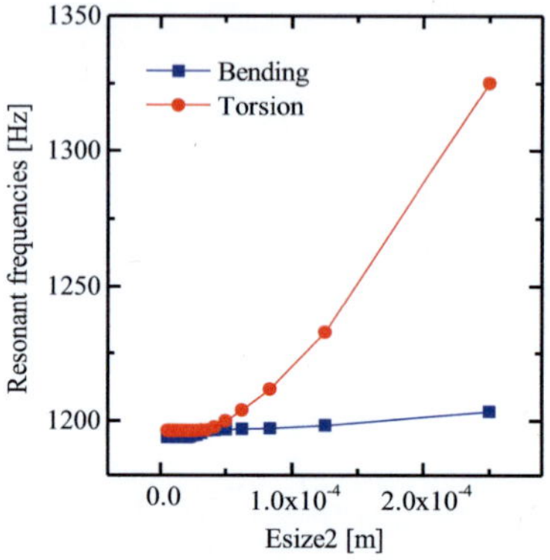

Fig. 3.22: MATLAB routine on *Esize2*, the bending and torsional resonant frequencies converge for element sizes smaller than *Esize2* = 35 µm.

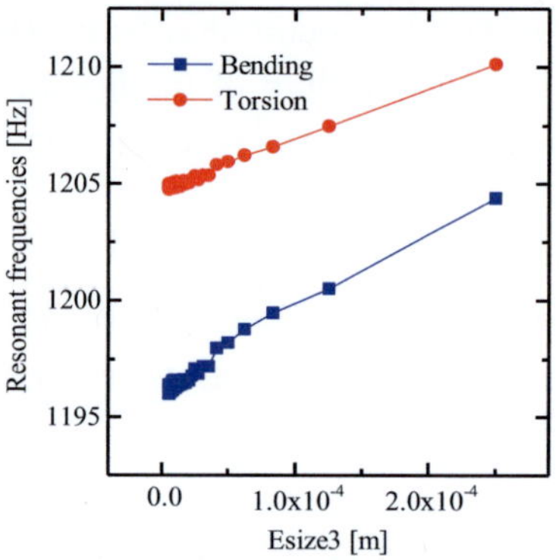

Fig. 3.23: MATLAB routine on *Esize3*, the bending and torsional resonant frequencies converge for element sizes smaller than *Esize3* = 7.5 µm.

After the previous studies on the three element sizes, a last MATLAB/ANSYS routine investigated the validity of *Esize1*. The results are still converging and the third of the thickness is a valid size. **Fig. 3.24** shows the final mesh of the sample.

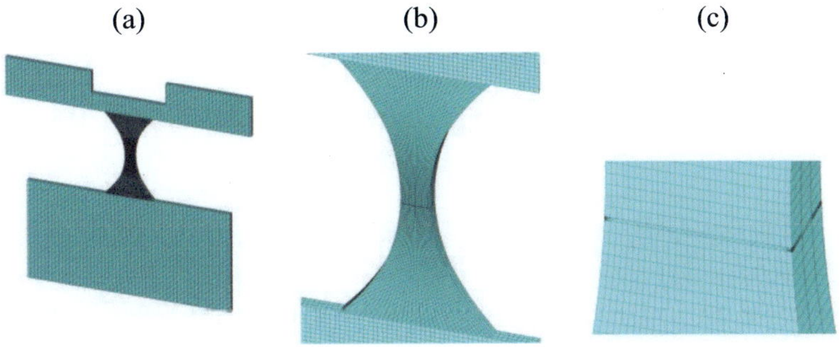

Fig. 3.24: Sample final mesh: (a) gripping section and mass with *Esize1*, (b) sample fillets with *Esize2*, and (c) refined volume with *Esize3*.

3.4.2 Crack geometry

The position of the crack has been determined with a static simulation of a model without crack, where the sample has been subjected to bending and torsional loading at the same time.

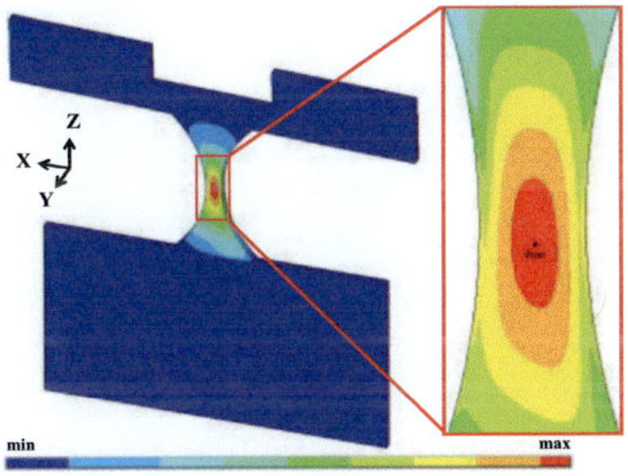

Fig. 3.25: Crack situated at the position of the maximal stress, determined by FE simulation.

The position of the maximal von Mises stress in the sample has been chosen as the center of the crack. The black dot (about 2.3 mm above the gripping section on **Fig. 3.25**) shows where the maximum is.

As a first step, only one crack has been simulated in the sample. The validation of the results for a simple model is important before implementing multiple cracks. Different crack models, such as block, prism and Anderson geometry, have been investigated. Nevertheless, only the Anderson model shows valid results, and is, therefore, described in this section. The Anderson model has a crack geometry inspired from the ANSYS example 'VM267: Inclined crack in 2D plate under uniform tension loading' (Source: ANSYS help). This example is one of the examples developed in Anderson's book: Fracture mechanics: fundamentals and applications [59]. The idea of this geometry is to divide the sample volume with an area representing the crack. The sample surface is separated by two areas, representing the crack opening. These two areas are moving and opening when the sample is under loading.

First, two blocks are created in the refined volume at the location of the highest stress. The crack is represented by the two coincident areas between the volumes V1 and V2, where the black dot is in **Fig. 3.26**.

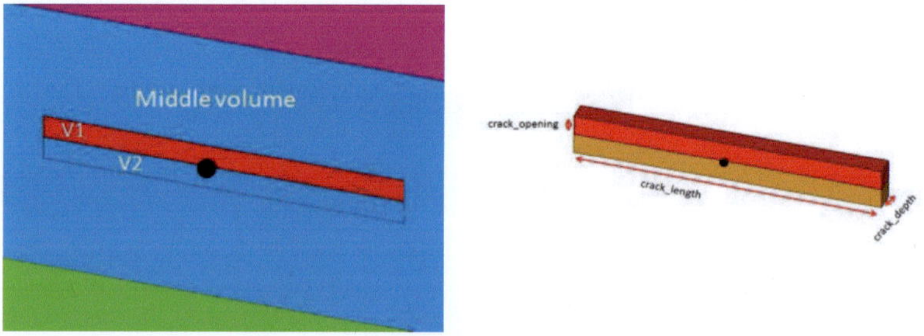

Fig. 3.26: Anderson geometry: creation of each crack flank with the volumes V1 and V2.

Then, eight cylinder quarters, V3 to V10 (**Fig. 3.27**) have to be created to allow for the crack opening and a structured fine mesh. The lines between the green point (**Fig. 3.27**b) and the cylinder center represent the crack tips. The double line between V1 and V2 represents the two crack edges. All other double points from the cylinder quarters are merged. If no quarters had been created, the merging would have glued the points and the crack opening would not have been possible.

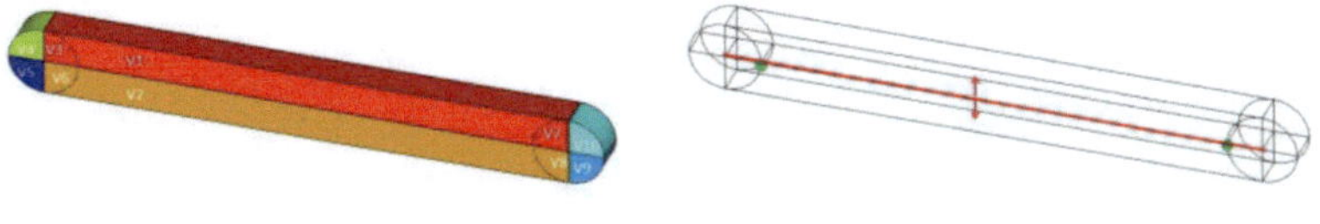

(a) (b)

Fig. 3.27: Anderson geometry: creation of cylinders for crack opening and for a structured fine mesh at the crack tip.

The mesh optimization from section 3.4.1 where the sample is divided into three parts with the element sizes *Esize1*, *Esize2* and *Esize3* is also used for the Anderson's model. Nevertheless the different volumes (V1 to V10), created for the crack implementation are meshed with an element size (*Esize4*) equal to 0.5 µm, insuring two elements through the cylinder radius.

3.4.3 Influence of the crack length on the resonant frequencies

Even though aluminum has been chosen to run the simulations; the analyses are reproducible for all materials. In this section, the influence of the parameters *crack_length* (**Fig. 3.26**) on the resonant frequencies is studied. In this case, *crack_depth* is set to 10 µm.

Fig. 3.28 shows the *crack_length* variation for the Anderson geometry. The decrease of the bending frequency is also equal to 0.02% and the decrease of the torsion to 0.01%. The complex construction of the geometry does not allow for having *crack_length* $\leq$ 2.5 µm. It explains the plateau at the beginning of the results.

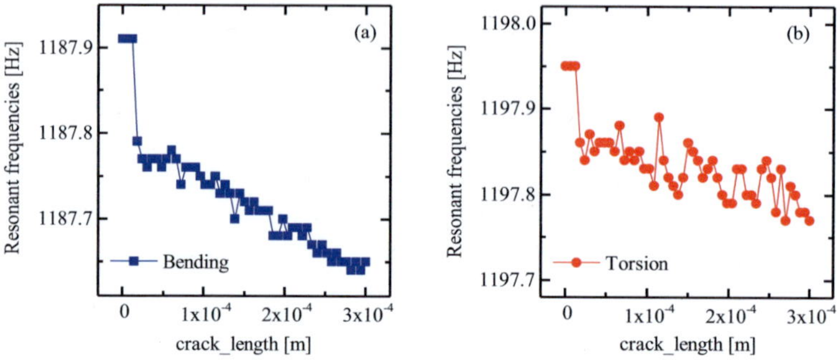

Fig. 3.28: Modal analysis: Anderson geometry: *crack_length* influence on the resonant frequencies.

In this case, the crack depth was fixed and the crack length was varying. In reality, the longer the crack, the deeper it grows. To perform a more realistic crack growth, it has been decided to correlate the length and the depth of the crack, according to microscopic observations after fatigue tests. It has been concluded that the depth of the crack is approximately ten times less than the length. The chosen correlation is: *crack_depth* = *crack_length* / 10.

With this first study, the Anderson crack geometry shows relevant results, with a reasonable calculation time.

3.4.4 Multiple crack implementation models

According to the previous study, the model with one crack built with the Anderson geometry and the associated element types and sizes is now considered as a reference. All the multiple crack simulations presented in this part are based on this model.

The crack geometry reference is detailed in **Fig. 3.29**a, and is defined by two parameters: *crack_opening* corresponding to the width of a volume (meshed with *Esize4*) where the crack is built **(Fig. 3.26)**, and *mesh_refine* corresponding to the half of the width of the volume (meshed with *Esize3*) where the crack model is implemented

(**Fig. 3.24**c). The meshing is only possible, if *mesh_refine* is equal to the *crack_opening* and an offset of *dist_trans*, where: *dist_trans* ≥ *mesh_refine* – 3/2 x *crack_opening*. This is the minimal distance where the mesh transition between *Esize4*, *Esize3* and *Esize2* is possible.

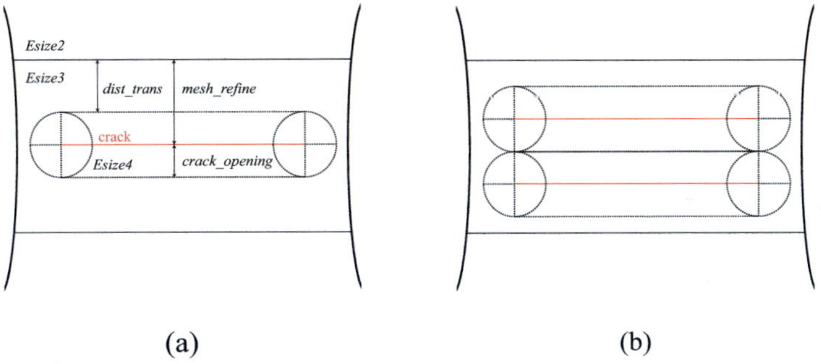

Fig. 3.29: (a) Anderson geometry reference: crack parameters. (b) Model: two cracks in one area.

The first implemented multiple crack model has two cracks (**Fig. 3.29**b). The geometry of the sample and of the crack are the same as previously explained. The only difference is that instead of having one crack in the refine volume (meshed with *Esize3*, **Fig. 3.24**c), two cracks have been implemented. To do so four volumes are forming the two cracks and each cylinder center is coincident to the tips of each crack (in comparison of the only two volumes from **Fig. 3.26**). The two cracks are equally placed inside the refined volume. The model parameter values are: *crack_opening* = 4.5 μm, *mesh_refine* = 10 μm. In this case, the distance of the mesh transition *dist_trans* is 3.25 μm.

The second implemented multiple crack model has three cracks (**Fig. 3.30**a). They are built in the refined volume (meshed with *Esize3*). The cracks are formed with six volumes. The middle crack is at the same position as the crack of the reference model and the two others are equally situated in the refined volume. The parameter values are: *crack_opening* = 4.5 μm, *mesh_refine* = 12 μm. The distance of the mesh transition is equal to *dist_trans* = 3 µm. In this case *dist_trans* = *mesh_refine* – 2 x *crack_opening*.

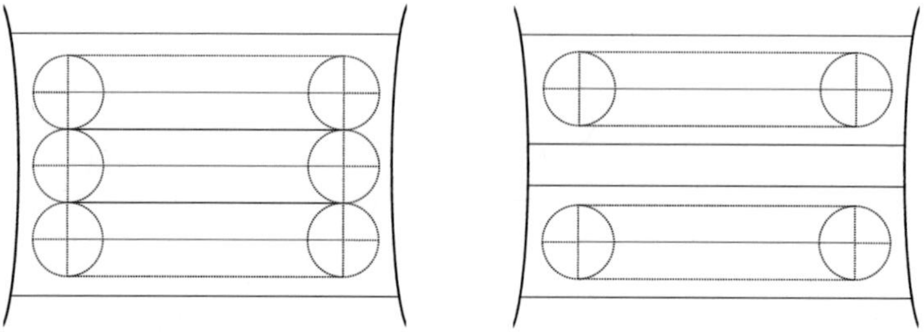

(a) (b)

Fig. 3.30: (a) Model: three cracks in one area. (b) Model: two cracks in two areas.

A third multiple crack model (**Fig. 3.30**b) has been performed with this time a second crack in another refined volume. This model has two refined volumes where the cracks are simulated. In each refined volume a crack is built with the same parameters as in the reference model (**Fig. 3.29**a). The distance between these two cracks, called *Ydist*, has no influence on the resonant frequencies, and is chosen equal to 100 µm. The parameter values are: *crack_opening* = 4.5 µm, and *mesh_refine* = 7 µm. The distance of the mesh transition is equal to 2.5 µm. In this case, *dist_trans* = *mesh_refine* - *crack_opening*. A similar structured mesh as the sample fillets of the reference model has been used for the volume between the two refined volumes.

In the fourth multiple crack model, two cracks are implemented in the same refined volume, but this time one on both side of the sample. The two cracks are symmetrical to the transversal plan of symmetry. The parameters are the same as the reference model (**Fig. 3.29**a).

This fifth model (**Fig. 3.31**a) is the same as the reference model but the crack orientation can be tilted. To allow for the rotation of the crack, the width of the refined volume has to be large enough to be able to have a crack orientation of *beta* = 90°. To avoid a too large refined volume, and thus too many elements, the maximal length of the crack has been chosen to be L = 50 µm. The parameter values are: *crack_opening* = 4.5 µm, *mesh_refine* = 27.5 µm. For a crack orientation of *beta* = 90° the distance of the mesh transition is equal to 2.5 µm (*dist_trans* = *mesh_refine* – $L/2$). In addition, for a crack orientation of *beta* = 0°, *dist_trans* is equal to 23 µm.

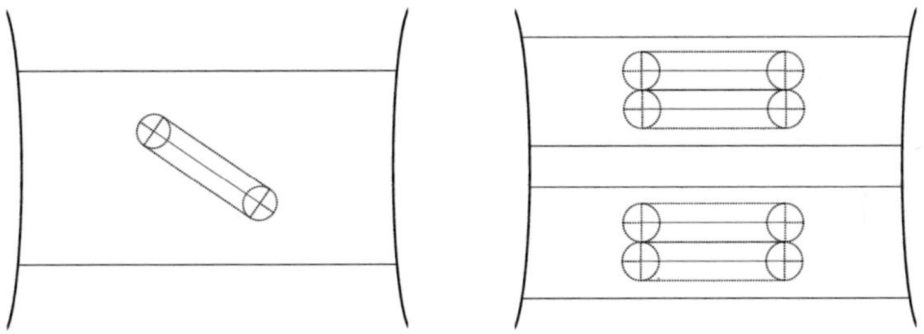

(a) (b)

Fig. 3.31: (a) Model: one crack in one tilted area. (b) Model: four cracks in two areas.

This sixth model (**Fig. 3.31**b) is based on the model for two cracks in one area and the model for two cracks in two areas. The model includes two cracks in each area. As for the model, two cracks in two areas the distance between the two cracked areas is arbitrary chosen equal to *Ydist* = 100 µm. The other parameter values are the same as for the model with two cracks in one area: *crack_opening* = 4.5 µm, *mesh_refine* = 10 µm. The distance of the mesh transition is equal to 3.25 µm. (*dist_trans* = *mesh_refine* – 3/2 x *crack_opening*).

3.4.5 Modal analyses

To understand the influence of the crack growth on the resonant frequencies, a MATLAB routine has been performed by starting a modal analysis with ANSYS in batch mode. To simulate the crack growth, the crack length has been increased between each step of the loop. At the first iteration, no crack is simulated, *crack_length* = 0 µm and the final step corresponds to *crack_length* = 300 µm. For each iteration on *crack_length*, the corresponding resonant frequencies of the first and second modes calculated by ANSYS are integrated in MATLAB. At the end of the loop, it is consequently possible to plot the sample resonant frequencies as a function of the crack length. This loop has been performed for all models previously presented.

Fig. 3.32 shows the resonant frequency variation as a function of the crack length for each model. **Fig. 3.32** represents the variation of the first resonant frequency corresponding to the bending out-of-plane mode and **Fig. 3.32**b is the second resonant frequency corresponding to the torsional mode.

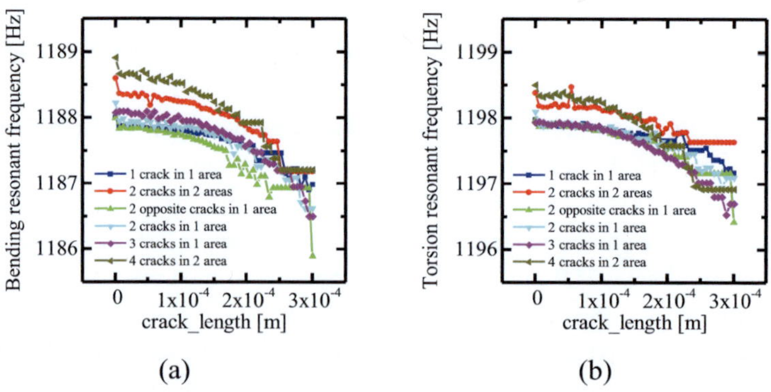

Fig. 3.32: Bending (a) and torsional (b) resonant frequencies of the different multiple crack models as a function of the crack length.

For a *crack_length* = 0 µm, no crack is implemented in the different models. Theoretically, the value of the resonant frequencies from all the models should be the same. However, it is possible to observe that the initial frequency values are not the same. This is due to the different geometric values of *crack_opening* and *mesh_refine* for each model. To be able to compare the different models, the data is normalized to a reference to have the same initial resonant frequency value.

An important change of the resonant frequencies appears at the first crack implementation compared to the initial value with no crack. Nevertheless, the decrease of the resonant frequencies, with growing crack length, is then homogeneous. The first value of the homogeneous decrease is an acceptable reference value. It corresponds to the third point of the frequency values of the reference model. The normalization equation is:

$$f_{model\ normalized} = \left(1 - \frac{\Delta f}{f_0}\right) * f_{model} \qquad \textit{Eq. 3.19}$$

Where

$$\frac{\Delta f}{f_0} = \frac{f_{reference\ model}(3) - f_{model}(3)}{f_{reference\ model}(3)} \qquad \textit{Eq. 3.20}$$

with $f_{reference\ model}$: the resonant frequency of the reference model, f_{model}: the resonant frequency of the model and $f_{model\ normalized}$: the normalized resonant frequency of the model.

After the normalization of the frequencies, the curves in **Fig. 3.33** have been obtained.

For both modes, the reference model, one crack in one area (dark blue) shows the least dependence on the crack length. The influence on the resonant frequencies gets stronger with the crack length (X axis in **Fig. 3.33**) and the crack number (different models).

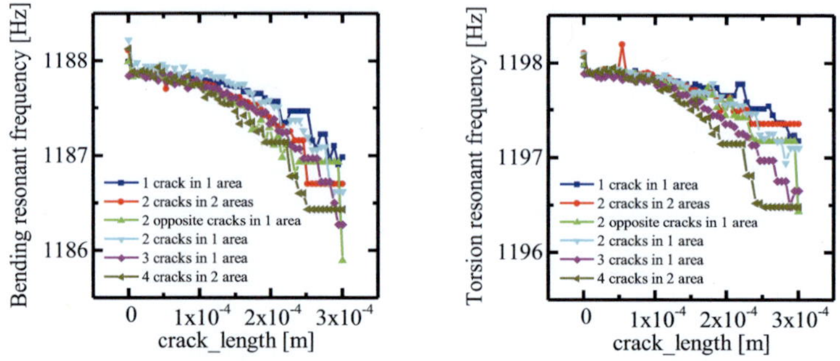

(a) (b)

Fig. 3.33: Bending (a) and torsional (b) normalized resonant frequencies of the different multiple crack models as a function of the crack length.

For the first mode (bending out-of-plane, **Fig. 3.33**a), the results of the models two cracks in one area (light blue curve) and two cracks in two areas (red curve) are similar. The final decrease of the resonant frequency of the model three cracks in one area (pink curve) is stronger than for the two previous models, and the decrease for the model four cracks in two areas (dark green curve) is even bigger. The higher the number of cracks on one side of the sample, the higher the decrease of the resonant frequency. For the model with two cracks on opposite sides (light green curve), the results are different. There are only two cracks but the decrease is almost as big as the one of the model with four cracks in two areas (dark green curve). The two cracks are positioned in the same transversal section; the stiffness in this section is therefore highly decreased by the growth of the two cracks.

Concerning the analysis of the second mode (torsion, **Fig. 3.33**b), the curves from the models with two cracks in one area (light blue curve), two cracks in two areas (red curve) and two cracks on opposite sides (light green curve) are close and have a stronger decrease than the decrease of the reference model. The decrease of the resonant frequency of the models with three cracks in one area (pink curve) and with four cracks in two areas (dark green curve) is even bigger. For torsion, the influence of the crack growth increases with the number of cracks, no matter where they are.

The influence of the crack growth is stronger for bending than for torsion. Indeed, the bending resonant frequency of the reference model has a final decrease of 1 Hz ($\Delta f/f_{initial}$ = -8.4 x 10^{-4}). While the torsion resonant frequency has a decrease of about 0.8 Hz ($\Delta f/f_{initial}$ = -6.7 x 10^{-4}). However, the number of cracks seems to have a stronger influence on the decrease of the resonant frequency for torsion. In both cases, it can be observed that the variation of the resonant frequencies seems to have a quadratic evolution.

The results obtained by these simulations are interesting and give new possibilities for the study of the influence of crack growth on the resonant frequencies. They are compared to the experimental results in the discussion at the end of section 6.1.2.

Further work could be done on the crack growth rate. Indeed, for these simulations, all cracks are growing at the same speed. It would be interesting to perform new simulations with the same models but with different crack growth rate. An even more interesting work would be the implementation of tenths of micro-cracks, as observed in experiments, and the analysis of their influence on the frequency decrease.

3.5 Influence of sample geometry on the resonant frequency

3.5.1 Difference between experimental and simulated results

The torsional and bending peaks of a measured frequency sweep for a Ni-sample are represented in **Fig. 3.34**. The bending and torsional amplitudes have maxima at 286 Hz and 293.5 Hz respectively. It should be noted that the frequencies of the two peaks are lower than the resonant frequencies calculated by FEM (**Fig. 3.11**). This difference can be explained by geometric imperfections due to the fabrication and preparation of the samples, as well as by damping (e.g. due to air drag), which is not included in the linear elastic FE model. The reduction of the gage section dimensions, during manufacturing or electropolishing, was identified as the most contributing factor due to the reduction in stiffness.

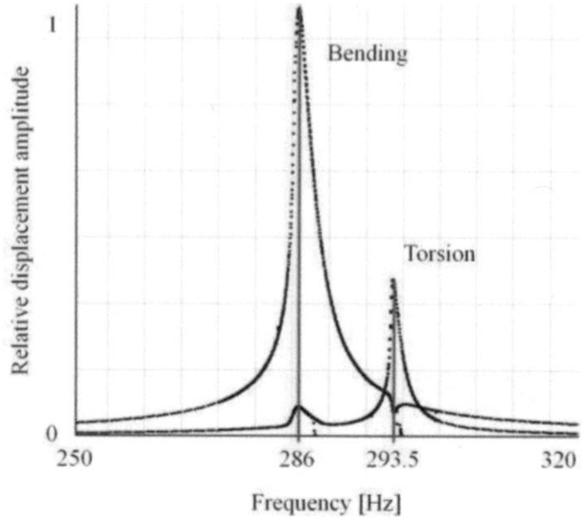

Fig. 3.34: Measured displacement amplitude for a frequency sweep showing two peaks, one for bending and one for torsion.

3.5.2 Frequency tuning by additional mass

Variations in manufacturing make it necessary to correct the mass distribution before the test to close the frequency gap between torsion and bending. Therefore, FE simulations were performed to find an optimally designed mass distribution in form of small added masses. Adding one drop of glue at the center of the mass leads to a decrease of the resonant bending frequency. If the drop is not placed at the center but shifted to one side of the mass, a mixture of a bending and torsional mode of the sample is observed. In case of two symmetrically applied drops, the resonant frequencies of both modes are affected. With increasing distance between the two drops from the center of the mass, the resonant bending frequency stays constant while the torsional one is reduced. A critical distance can be found, for which the two frequencies are equal. Finding this critical distance by simulation and recreating it in an experimental setup are completely different matters. In an attempt to bring the torsional and bending modes to the same frequency, super glue and lead balls were added with the largest distance from the center of the mass, at the end of the mass wings. Most trials resulted in a reduction of both resonant frequencies, and also tended to bring the modes much closer, but it is very difficult to apply the masses with the accuracy required to attain matching torsional and bending resonant frequencies. Although it is very difficult to attain matched modes, the addition of masses does noticeably increase the measured torsional amplitudes, allowing for an almost perfect multiaxial condition. Finally, as a quantified example, placing a 5 mg lead ball at the middle of the mass leads to a decrease of 12 Hz of the bending resonant frequency. Using a precision balance with a 0.1 mg resolution allows for a controlled minimum resonant frequency change of 0.24 Hz. Adding more weight does not change the mass distribution in the sample measurably. Furthermore, as the overall geometry stays the same, the stress distribution in the gage section is not influenced. This was also evaluated by FE simulations.

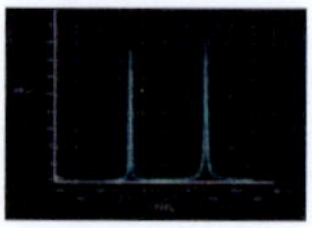	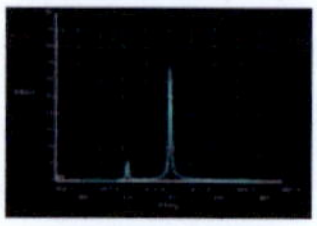	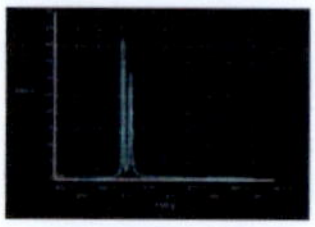	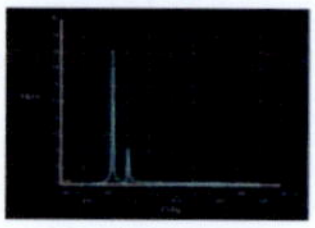
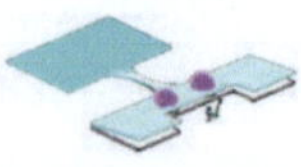		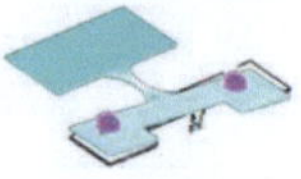	
1st mode bending 471 Hz	1st mode bending 471 Hz	1st mode torsion 470 Hz	1st mode torsion 468 Hz
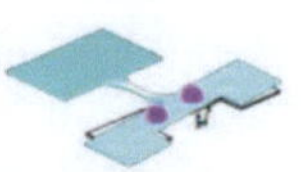	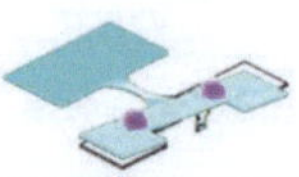		
2nd mode torsion 479 Hz	2nd mode torsion 475 Hz	2nd mode bending 471 Hz	2nd mode bending 470 Hz

Tab. 3.3: Influence of two added lead balls on the sample resonant frequencies.

4 Experimental methods

This chapter gives details on the sample preparation and characterization. The samples were made out of foils using a laser cutting or Electrical Discharge Machining (EDM) process. They were then electropolished in order to minimize the role of surface defects. Characterization methods such as 3D laser scanning microscopy, Scanning Electron Microscopy (SEM) and Focused Ion Beam (FIB) Microscopy are presented.

4.1 Sample preparation

4.1.1 Material

Aluminum (99.999%), annealed nickel (99.00%), and OFHC copper (99.95%) samples, acquired from thin metal foils (Goodfellow, United Kingdom), were used for preparing the samples. **Tab. 3.1** summarizes the foil properties. The hourglass shaped samples, having a width of 500 μm, were cut from Al thin foils with a thickness of 250 μm, and from Cu thin foils with a thickness of 200 μm, while the Ni-samples had a short straight section (about 400 μm) with a 500 x 200 μm² cross section.

Ref.	Material	Thickness [mm]	Purity	Temper	Tensile strength [MPa]	Yield strength [MPa]
AL000630	Al	0.25	99.999%	As rolled	130-195	110-170
CV000670	Cu	0.20	99.95%	Hard	314	270
NI000550	Ni	0.20	99.0%	Annealed	660	480

Tab. 4.1: Material properties.

4.1.2 Sample cutting

The Ni-samples were cut with a laser process (Bergmaier GmbH, Bruchsal, Germany) and by Mr. Besser at the KIT, while the Al and Cu were cut by EDM (Bergmaier GmbH).

No information was given on the cutting processes. The sample sides were nevertheless controlled after cutting in order to ensure a reasonable quality. The size of the zone, where the microstructure could have been influenced due to high temperature, was small enough to be removed after electropolishing.

4.1.3 Electropolishing

(a) (b)

Fig. 4.1: Ni-sample (sample 0) before electro polishing (a) and after (b).

In a consecutive process, all samples were electropolished in the gage section in order to remove thermally affected zones from manufacturing and to improve the surface quality in order to minimize the influence of surface defects on the damage formation. An automatic, microprocessor controlled electrolytic polishing machine (LectroPol-5, Struers, Denmark) has been used with the following parameters.

Material	Al	Cu	Ni
Electrolyte	A2	D2	A2
Voltage [V]	40	25	40
Flow rate	10	14	14
Time [s]	5	10	12

Tab. 4.2: Electropolishing parameters.

The surface roughness was measured before and after electropolishing with a 3D laser scanning microscope (VK-9700, Keyence, Japan, see section 4.2.1). **Tab. 4.3** is an example of a Ni-sample roughness (Japanese standard JIS B 0601:1994).

	Before	**After**
***Rz* (µm)**	3.125	1.844
***Ra* (µm)**	0.164	0.068

Tab. 4.3: Surface roughness of a Ni-sample before and after electropolishing.

The roughness on all types of materials was checked for a couple of samples. After confirmation of the obtained roughness and of the sample surface high quality, the same electropolishing process has been applied to all samples.

Manufacturing imperfections resulted in variations in dimensions; thus, each sample thickness was measured individually. Al thicknesses are listed in **Tab. 8.4** and **Tab. 8.5**, while Cu thicknesses are summarized in **Tab. 8.6**, **Tab. 8.7** and **Tab. 8.8** (see Appendix).

4.2 Microstructural analysis

4.2.1 3D laser scanning microscope

After electropolishing, a micrograph of each sample has been taken with a Keyence VK 9700 3D laser scanning microscope. With this microscope, the original state of each sample has been observed before fatigue. In the same way, a micrograph of each sample has been taken after a fatigue test.

This microscopy method is similar to a Confocal Laser Scanning Microscopy. It takes high resolution micrographs with a height regulation. It is able to focus on different heights, allowing for a 2.5D optical image reconstruction. A 50x objective and a 0.01 μm height step were used for image acquisition, giving information on the sample surface topography in the nm regime.

4.2.2 Scanning Electron Microscopy and Focused Ion Beam Microscopy

After a fatigue test, the sample microstructure has been investigated with a dual beam microscope (see chapter 5 Results). Surface roughness, cracks on the surface or even cracks through the thickness of the sample, as well as any kind of defects can be observed. A Nova 200 Nanolab, (FEI, USA) has been used. It includes an SEM and a FIB in the same device.

To observe the microstructure on the sample surface, micrographs were taken using the SEM. A 10 kV accelerating voltage for the electron beam was used with an Everhart-Thornley Detector (ETD) for low magnification or a Through Lens Detector (TLD) for high magnification.

Finally, to look at subsurface defects (such as micro-cracks or pores), cross sections were cut into the sample using the Gallium beam of the FIB. A Platinum layer is first deposited on the sample surface to protect it. A staircase cross section is then milled, and the obtained surface is finally cleaned with some low current to improve the quality of the cross section.

5 Results

The experimental results are presented in this chapter. In a first step, the damage detection method is examined. Later, the results are presented from the investigation of early damage evolution as a function of the relative frequency decrease using nickel and copper samples. Finally, evidence is shown displaying crack initiation for fcc materials in the HCF and VHCF regimes with Al and Cu samples.

5.1 Damage detection method

Bending experiments on Ni and Cu micro-samples were conducted in the HCF regime to validate the experimental resolution of the setup, as well as the damage detection method.

Due to the accumulating fatigue damage, the resonant frequency of the bending micro-samples decreases with increasing cycle number. In **Fig. 5.1**, the relative resonant frequency change $\Delta f / f_{\text{initial}}$ versus cycle number of a micro-sample during a typical HCF test is shown. Up to $N = 10^5$ cycles, the resonant frequency is only slightly changing by roughly $\Delta f / f_{\text{initial}} = -1 \times 10^{-4}$. In this regime, extrusions develop in many grains and consecutively micro-cracks initiate in single grains. Beyond 10^5 cycles the relative frequency change decreases below about $\Delta f / f_{\text{initial}} = -1 \times 10^{-3}$ and micro-cracks propagate through grain boundaries to form longer cracks. After a relative frequency decrease beyond roughly $\Delta f / f_{\text{initial}} = -5 \times 10^{-3}$ a long crack propagates through the sample changing the resonant frequency at a fast rate.

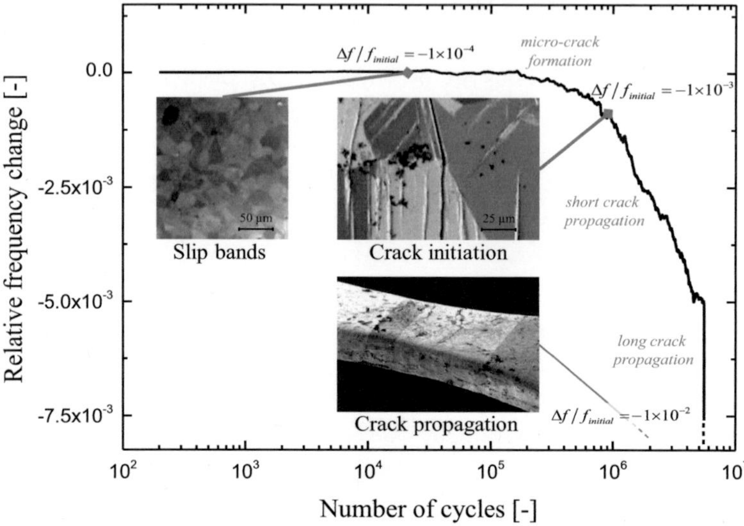

Fig. 5.1: Relative resonant frequency shift versus cycle number during an HCF test compared to damage in a Ni-sample.

5.1.1 First damage: slip bands

The first slip bands were accompanied by a decrease of the relative resonant frequency on the order of roughly $\Delta f / f_{\text{initial}} = -1 \text{ x } 10^{-4}$ (**Fig. 5.2**). At this stage, FIB cross sections showed that no cracks were observable. **Fig. 5.2**a is a laser scanning micrograph of the sample surface, where each dark line represents a slip band. The laser scanning microscopy allows for tracking the slip band height formation quantitatively but the technique is limited in spatial resolution due to the used wavelength, in this case the height of the smallest slip bands cannot be evaluated. **Fig. 5.2**b is an ion-enhanced electron micrograph, it shows the grain morphology of the sample surface with an intense grain orientation contrast. Each dark line (highest ones indicated by white arrows) represents a slip band constrained within the grain boundaries.

At this stage, the slip bands are typically constrained to single grains, though several similarly oriented neighboring grains show slip bands crossing their grain boundaries. Micro-cracks could not be found at this stage of fatigue damage.

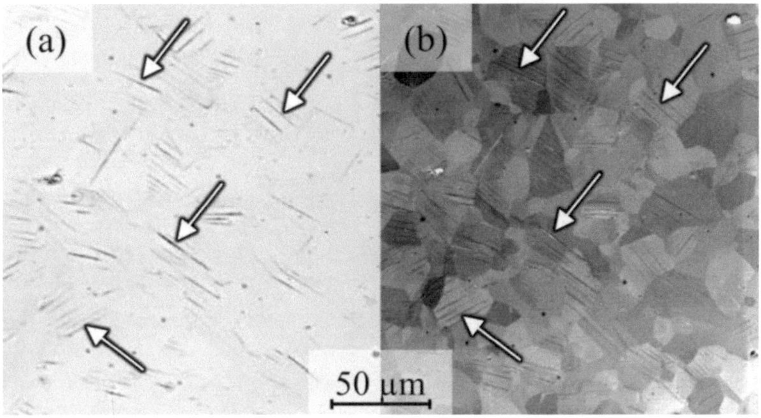

Fig. 5.2: Laser scanning micrograph (a) and an ion-enhanced electron micrograph (b) of slip band formation at the same location after a relative resonant frequency decrease of $\Delta f/f_{initial} = -1 \times 10^{-4}$ on a Ni-sample.

5.1.2 Second phase: crack initiation

In the next stage, the extrusions and intrusions are increased in their amplitude, leading to crack initiation. The micro-sample showed a resonant frequency decrease of $\Delta f/f_{\text{initial}} = -1 \times 10^{-3}$ corresponding to multiple crack initiations at the surface. **Fig. 5.3**a shows a scanning electron micrograph of typical fatigue damage of a Cu grain. In this case, a Cu-sample has been chosen for its precise representative results. Ten such fatigued grains out of about two thousand have been observed on the upper sample surface. The red line represents where the sample has been cut as seen in **Fig. 5.3**b. The focused ion beam cross section shows the initiated damage growing from the surface and penetrating into the depth.

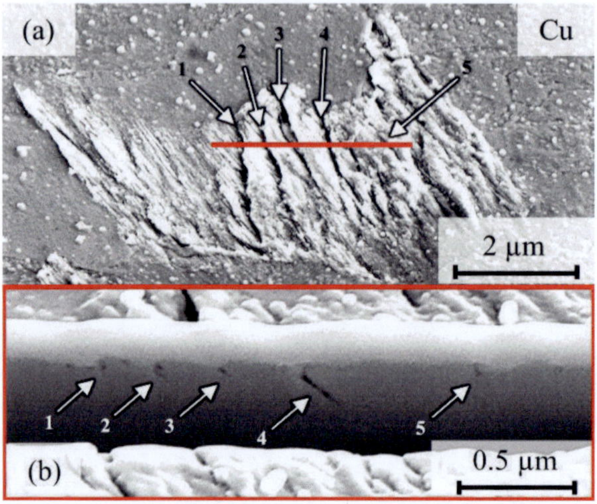

Fig. 5.3: Cu-sample: (a) Scanning electron micrograph of a fatigued grain after a relative resonant frequency decrease of $\Delta f / f_{initial} = -1 \times 10^{-3}$, (b) scanning electron micrograph of a FIB cross section at the red line in (a) showing the micro-crack formation at extrusions.

After a relative decrease of $\Delta f / f_{\text{initial}} = -1 \times 10^{-3}$, the fatigue-induced surface roughening has increased dramatically, as can be seen in **Fig. 5.4**b for a Ni micro-sample. **Fig. 5.4**a illustrates the 'as prepared' sample and its electropolished surface. **Fig. 5.4**b shows the sample after $N = 10^6$ cycles under a bending stress amplitude of $\sigma_a = 200$ MPa. Both images were acquired with a laser scanning microscope. Extrusions and cracks appeared at the surface, where the maximum stress amplitude is expected from FE simulations (**Fig. 5.4**c). At this stage, several micro-cracks can be observed at the sample surface. **Fig. 5.4**d is a magnification of **Fig. 5.4**b and shows the damage in the straight section. A further magnification (**Fig. 5.4**e) shows that the extrusions (white lines) are covered by Ni oxide, while the crack (dark line) follows a path along extrusions and crosses a grain boundary into the neighboring grain (see arrows). The orientation of the crack is changing at the grain boundary.

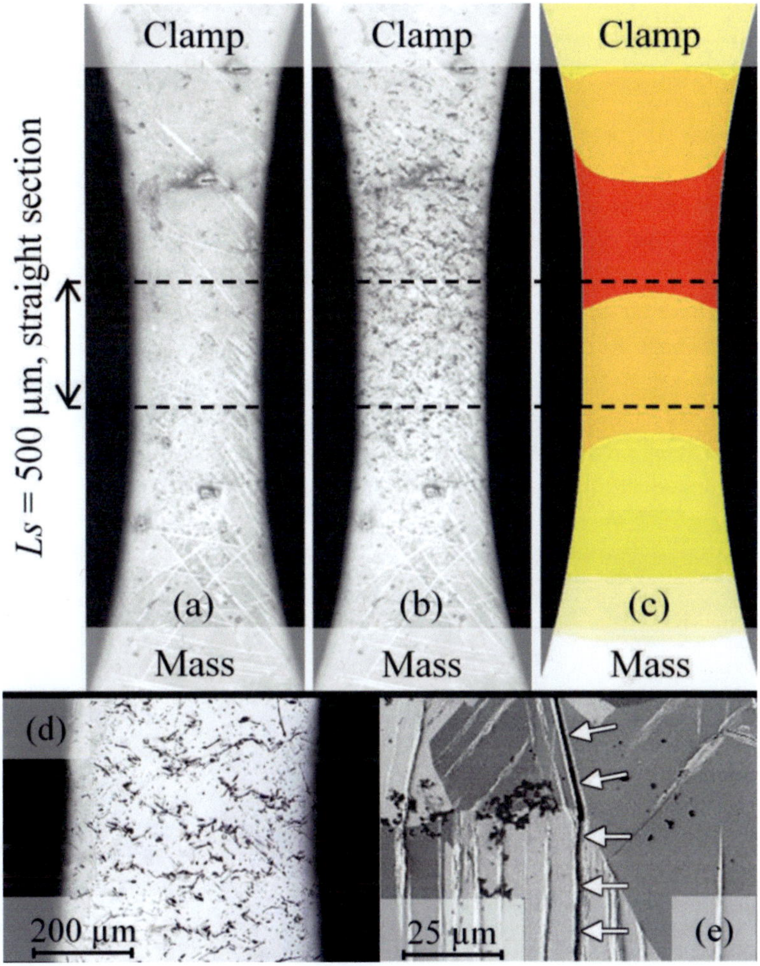

Fig. 5.4: Laser scanning micrographs: (a) Ni micro-sample before fatigue, (b) after $N = 10^6$ cycles with $\sigma_a = 200$ MPa maximal stress amplitude. (c) Stress distribution according to FE simulation. Red to white corresponds to high to low stresses. (d) Laser scanning micrograph, (e) scanning electron micrograph of the crack initiation after a relative resonant frequency decrease of $\Delta f / f_{initial} = -1 \text{ x } 10^{-3}$.

5.1.3 Third phase: crack propagation

Finally, at a relative resonant frequency decrease of $\Delta f / f_{\text{initial}} = -1 \times 10^{-2}$ one of the micro-cracks propagates through the sample. **Fig. 5.5**a shows multiple cracks (white arrows) at the sample surface going through the width and the thickness. The FIB cross section in **Fig. 5.5**b documents how the crack propagates through the grains and crosses grain boundaries. At this stage, the sample is considered to have failed. It is not possible anymore to track the phase angle between the actuation and the sample response. Here, the damage formation and the associated frequency shift are proceeding too fast for the experimental control to adjust the resonance frequency. Furthermore, the resonant mode becomes unstable at some crack length due to the severe change in geometry. In this case the test is stopped and the sample is marked as end-of-life.

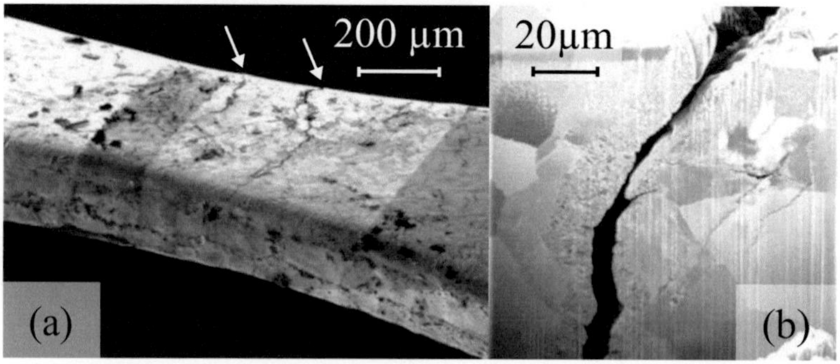

Fig. 5.5: (a) Scanning electron micrograph. Cracks emphasized by the white arrows. (b) Scanning electron micrograph of a FIB cross section showing the crack propagation after a relative resonant frequency decrease of $\Delta f / f_{initial} = -1 \times 10^{-2}$.

Fig. 5.6 shows an ion-enhanced scanning electron micrograph of a FIB cross section, an intragranular crack propagation after a relative resonant frequency decrease of $\Delta f / f_{\text{initial}} = -1 \times 10^{-2}$ can be observed. Moreover, another interesting fact is observed on several samples. One of them is shown in **Fig. 5.6**, the white arrows emphasize a thin layer of a couple of hundreds nm at the sample surface. The different contrast compared to the grain, on which the thin layer is seen, corresponds to a different crystallographic microstructure.

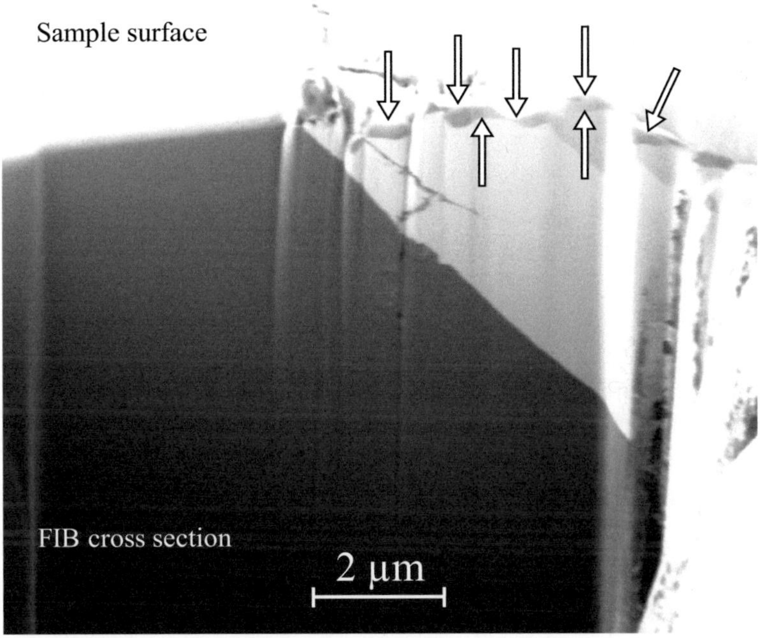

Fig. 5.6: Ion-enhanced scanning electron micrograph of a FIB cross section showing the crack propagation after a relative resonant frequency decrease of $\Delta f / f_{initial} = -1 \times 10^{-2}$. A intragranular crack can be observed. The white arrows mark a thin layer of a couple of hundreds nm.

5.2 Investigation of the early damage evolution

5.2.1 Correlation between the amount of damage and the relative frequency decrease

A further study has been done with the sample Cu44 where the early damage evolution has been observed. The amount of damage has been optically estimated by calculating the dark pixel amount on the sample surface, and has then been quantitatively correlated to the relative frequency decrease $\Delta f/f_{initial}$.

Tab. 8.1 (see Appendix) enumerates the different tests done with Cu44. Each test has been stopped after a certain relative frequency decrease. The steps have been chosen such to be able to observe the damage initiation in a single grain, as well as to follow the propagation through different grains.

To correlate the amount of damage and the frequency change, an optical micrograph has been taken after each test (**Fig. 5.7** - **Fig. 5.10**).

Each micrograph has been tilted and translated to be at the exact same position. It has also been corrected to have the same contrast, and luminosity. A grayscale mode has finally been activated to convert the image with 256 levels of gray, from 0 (black) to 255 (white).

The damage evolution can be observed in the following micrographs. The extrusions grow first in single grains from test #1 to test #15. They then start to propagate in the next grains from test #16 to test #20, and finally covers a high percentage of the sample surface until test #37.

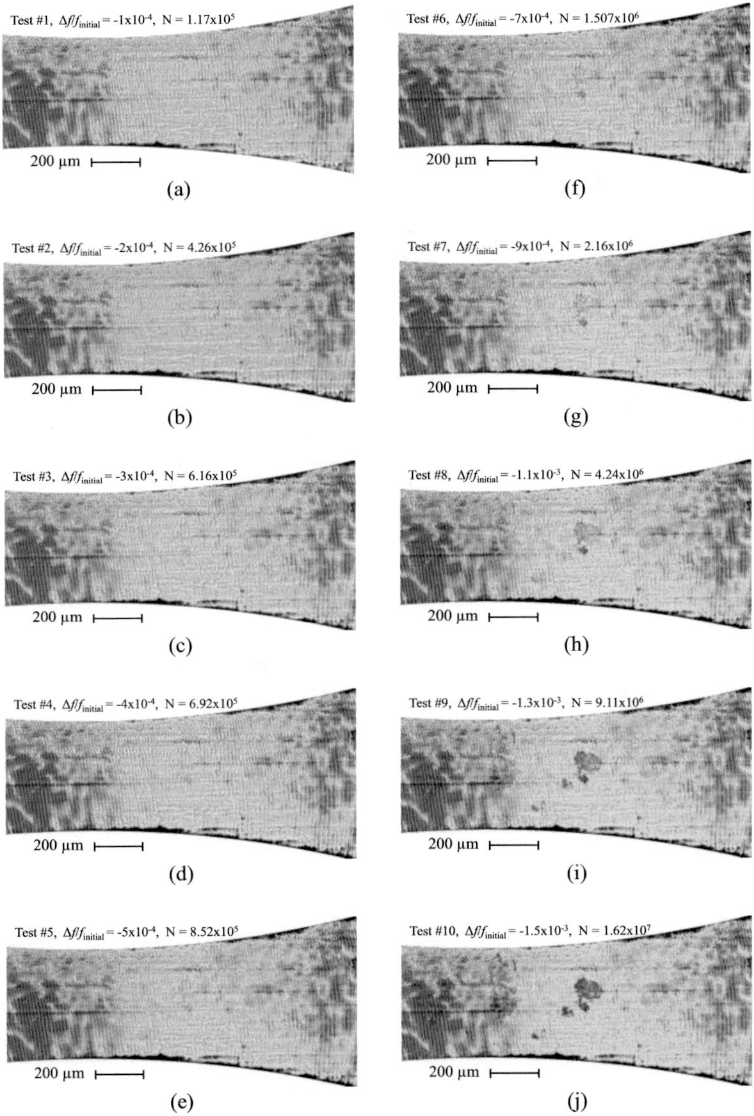

Fig. 5.7: Cu44: Optical micrographs of the sample surface from test #1 (a) to test #10 (j) where damage evolution and relative resonant frequency change can be correlated.

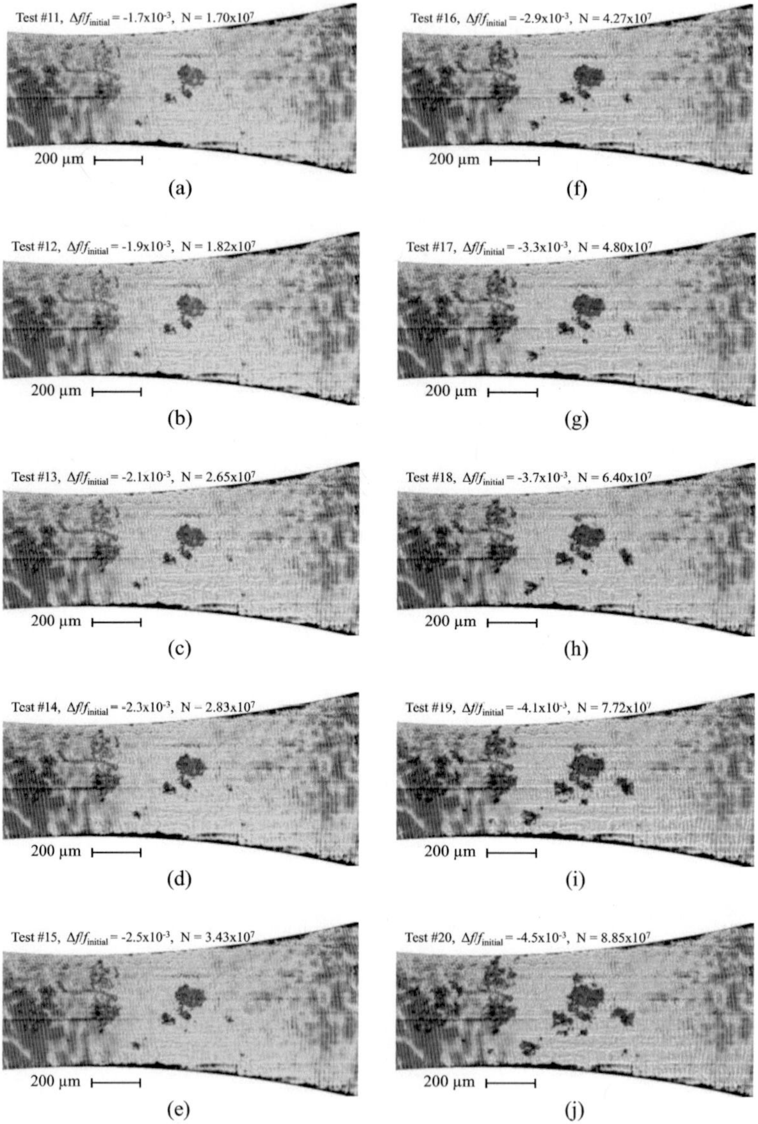

Fig. 5.8: Cu44: Optical micrographs of the sample surface from test #11 (a) to test #20 (j) where damage evolution and relative resonant frequency change can be correlated.

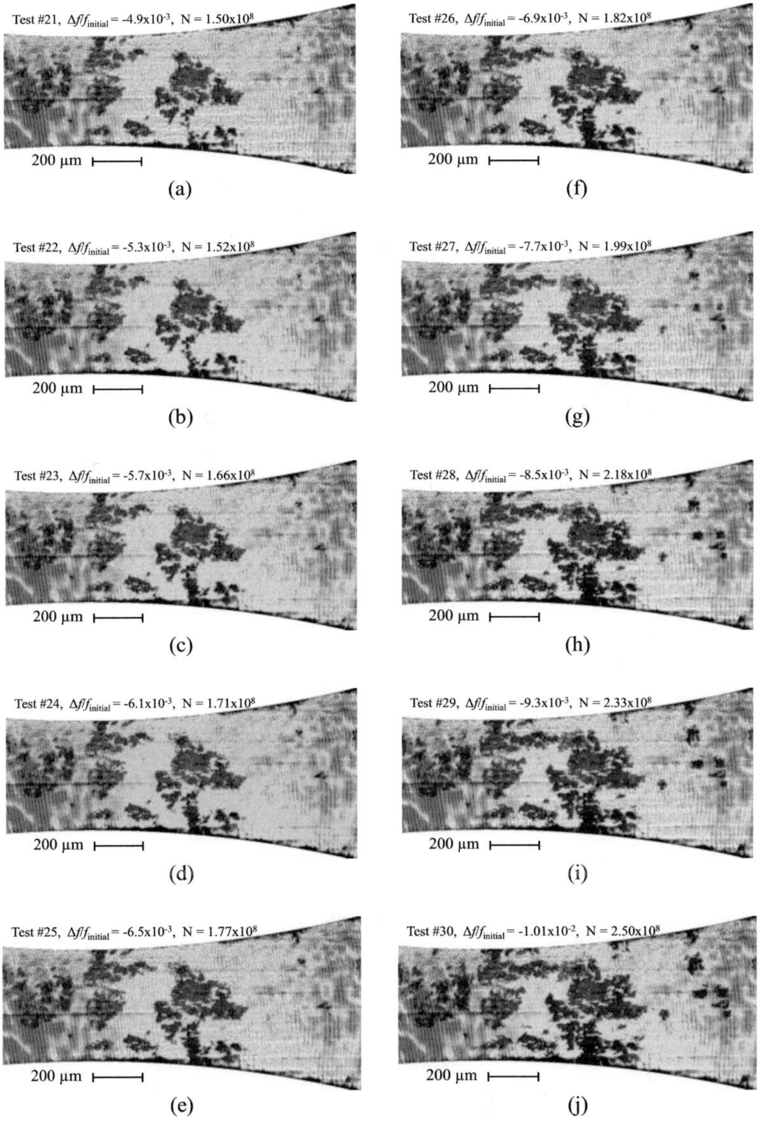

Fig. 5.9: Cu44: Optical micrographs of the sample surface from test #21 (a) to test #30 (j) where damage evolution and relative resonant frequency change can be correlated.

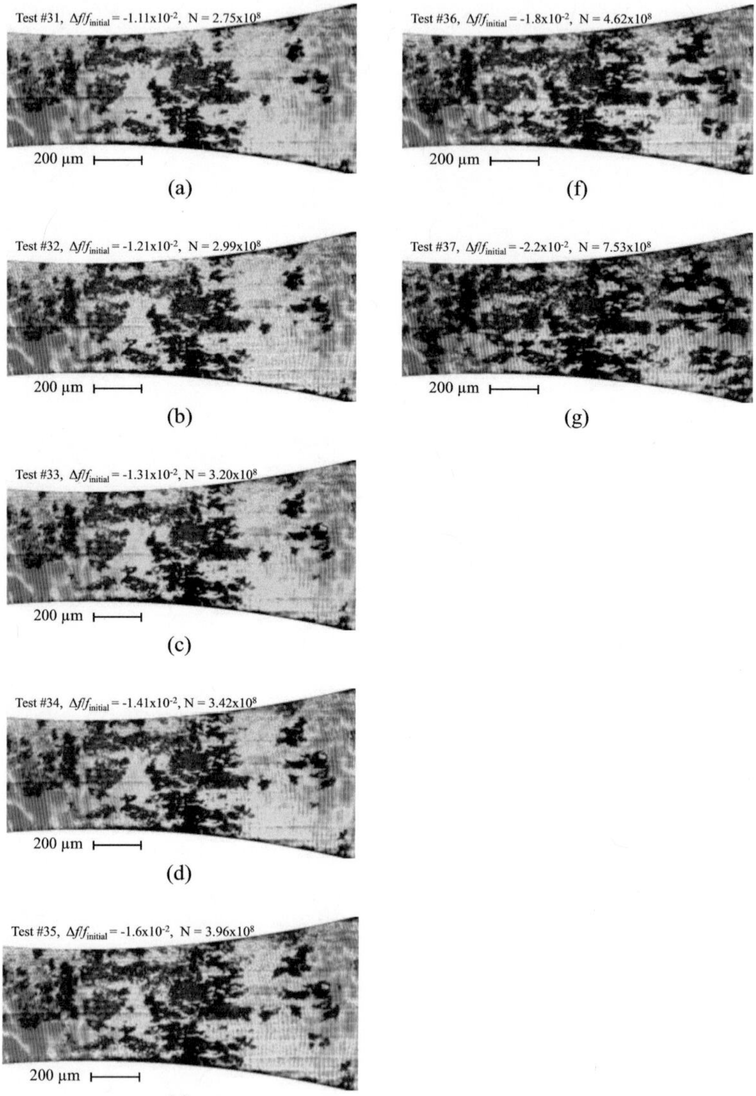

Fig. 5.10: Cu44: Optical micrographs of the sample surface from test #31 (a) to test #37 (g) where damage evolution and relative resonant frequency change can be correlated.

The evolution of damage has been observed and followed with the previous micrographs. It is now also possible to quantify the amount of damage by calculating the percentage of pixel in the micrograph below a certain level of gray. **Tab. 8.2** (see Appendix) lists the estimation of the sample damage amount by looking at the dark pixels.

Fig. 5.11 shows different plots of this estimation. The trend is the same for the three different gray level ranges. It can be observed that the increase of the relative amount of dark pixels as a function of the relative decrease of the resonant frequency is linear (**Fig. 5.11**a). It can also be seen in **Fig. 5.11**b that damage appears right at the beginning, needs a long time to develop and covers the entire surface at the end.

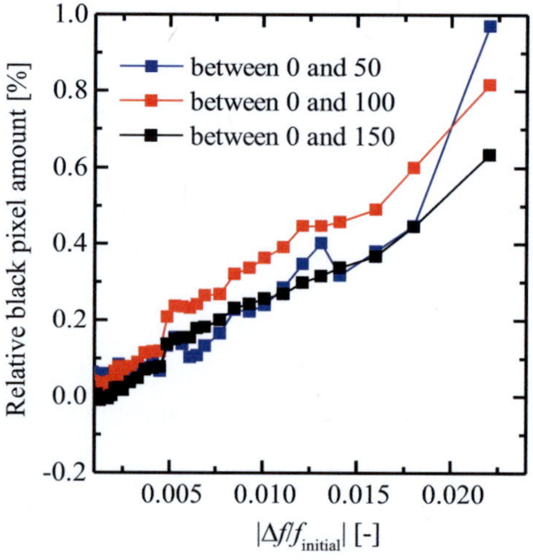

(a)

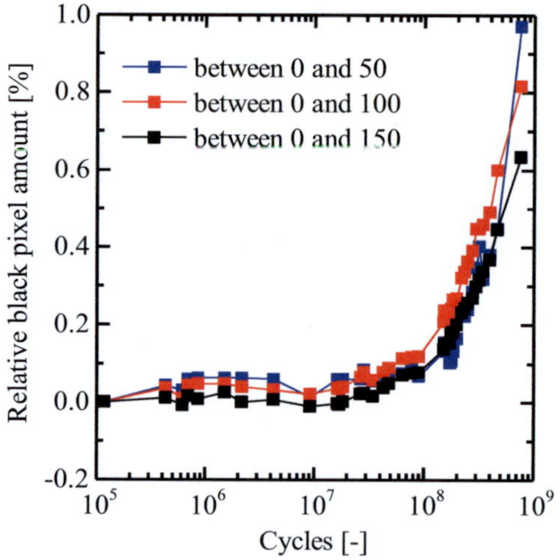

(b)

Fig. 5.11: Relative pixel amount between certain gray levels as function of the relative frequency decrease (a) and the number of cycles (b).

The same method can be used to look at the damage evolution in a single grain. **Fig. 5.12** shows the damage evolution in the grain where the damage first initiated.

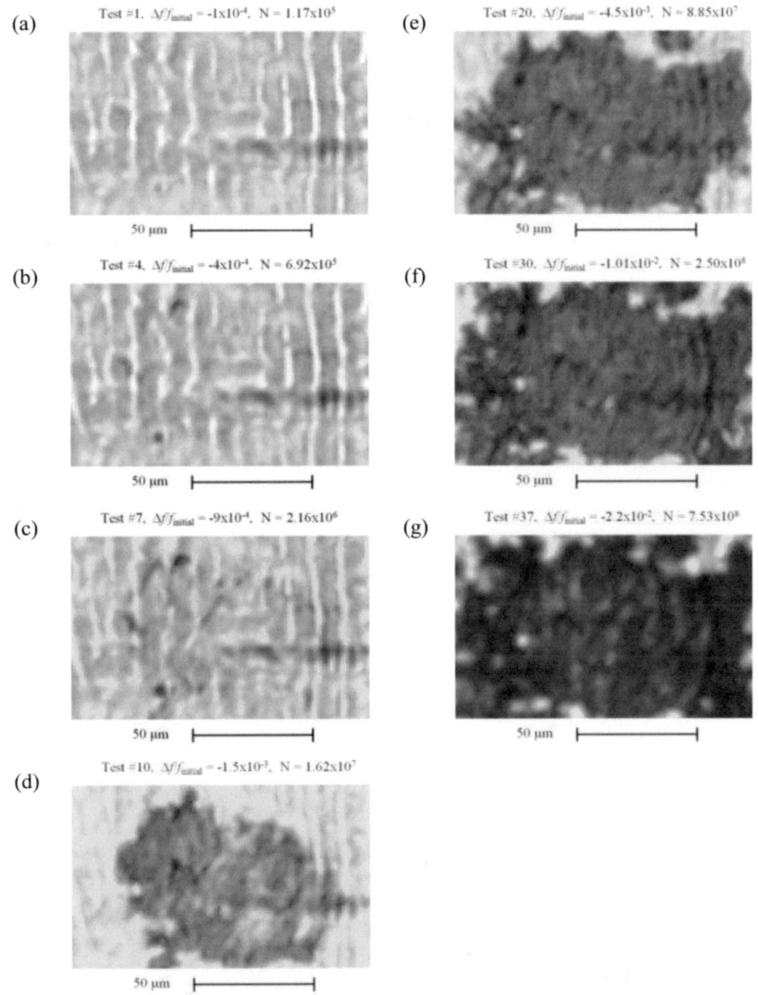

Fig. 5.12: Cu44: Optical micrographs of a single grain from test #1 (a) to test #37 (g) where damage and frequency change can be correlated.

The amount of damage can be quantitatively estimated (**Tab. 8.3**, see Appendix) by looking at the amount of dark pixels on the surface (same method as for **Tab. 8.2**).

Fig. 5.13 illustrates this estimation. From the red curve, corresponding to the 0 to 100 gray level range, it can be observed that the damage evolution is not linear as a function of the relative decrease of the resonant frequency (**Fig. 5.13**a). It increases faster than on the whole sample (**Fig. 5.11**a). This grain, where damage initiated, looks almost fully damaged after $\Delta f/f_{\text{initial}} = 0.01$. Though, the relative black pixel amount as a function of the number of cycles (**Fig. 5.13**b) shows the same trend as for the whole sample (**Fig. 5.11**b).

The validity of the black and blue curves for the relative black pixel amount on this grain is discussed in section 6.2.1.

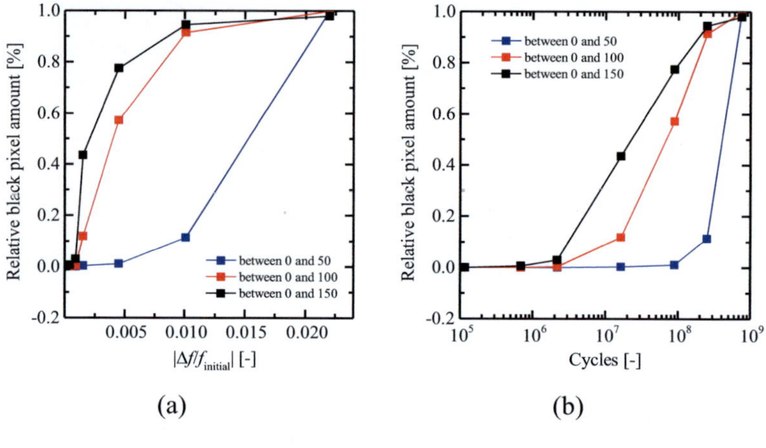

Fig. 5.13: Relative pixel amount between certain gray levels for a single grain as function of the relative frequency decrease (a) and the number of cycles (b).

5.2.2 Outlook: in-situ method coupled with EBSD and DDD

The previous optical measurement of the dark pixel amount has been used to establish the correlation of the amount of damage on the sample surface with the relative frequency decrease $\Delta f/f_{initial}$. For each optical micrograph, the test has been stopped, and the sample surface has been observed under a microscope. This time consuming method could be improved by using an optical microscope built in the setup, looking at the sample while it is fatigued. It would be a large gain of time and each sample would be tested in-situ.

Some Electron Backscatter Diffraction (EBSD) analyses could also be systematically done before testing in order to determine the grain orientation on the sample surface (e.g. for a Ni-sample **Fig. 5.14**).

(a) (b)

Fig. 5.14: Ni9: (a) Ion-enhanced scanning electron micrograph, (b) MapZ of an EBSD scan.

Comparing these EBSD results with the damage location (e.g. the first damaged grains where cracks are initiated), allows for finding the weakest grain and so the weakest orientation. This orientation could be used as an input for DDD simulations. The DDD allows for the simulation of the dislocation network and the calculation of the dissipated energy, as well as the dislocation density and the number of dislocation reactions. These simulations are also a good tool to understand the possible damage formation in the VHCF regime. The work of Dr. Daniel Weygand and his group on this topic is shortly described in section 6.2.2.1.

5.3 Investigation of the crack initiation in fcc materials

After the description of the damage detection method (5.1) and of the early damage evolution (5.2), this section presents the investigation of the crack initiation for Al and Cu micro-samples.

5.3.1 Critical stress amplitude

First, the resonant frequency behavior of the sample depending on the testing stress amplitude has to be pointed out. **Fig. 5.15** represents the relative resonant frequency change of a sample as function of the number of cycles during a VHCF test. The two different curves correspond to two different tests (of two different samples) with different stress amplitudes. It can be seen that the resonant frequency behavior changes depending on the stress amplitude of the test.

The curve on the lower part of **Fig. 5.15** shows a decrease of the relative resonant frequency along the number of cycles, where the stress amplitude is considered as overcritical. In this case, the decrease of the relative resonant frequency correlates with the observation of damage evolution, such as slip band formation, crack initiation up to the crack propagation.

Another behavior was observed when the stress amplitude is below a certain threshold, named for this work as 'critical amplitude'. In this situation, when the amplitude is subcritical, the relative resonant frequency is increasing with increasing number of cycles (top curve in

Fig. 5.15). One can observe a smooth increase from the beginning on, up to about $\Delta f / f_{\text{initial}} = 5 \times 10^{-3}$ when the test has been stopped.

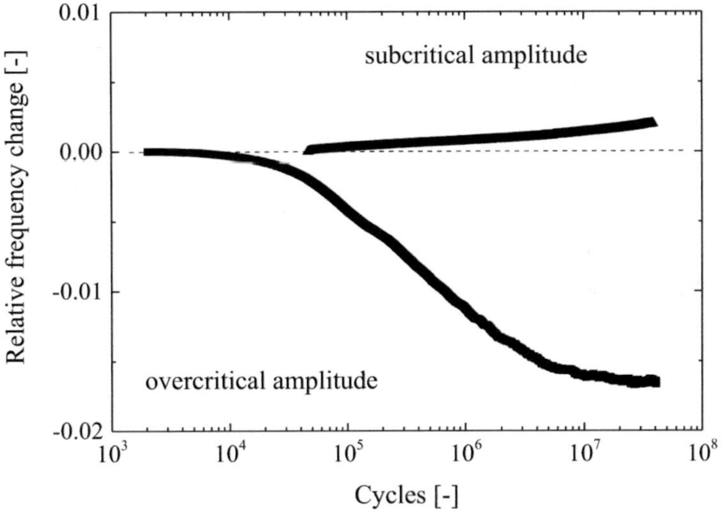

Fig. 5.15: Critical stress amplitude: a decrease of the resonant frequency appears when the stress amplitude is overcritical, on the contrary an increase of the sample resonant frequency occurs during a test with a subcritical stress amplitude.

5.3.2 Aluminum fatigue tests

The following section presents the results of fatigue tests established with Al samples.

5.3.2.1 Lifetime until $|\Delta f/f_{initial}| = 2 \times 10^{-3}$

Fig. 5.16 summarizes the Al test results, as on S-N curve, with the lifetime criterion explained in 5.1.2. It does not correspond to the lifetime of failure, but the lifetime until a change of the resonant frequency of $|\Delta f / f_{\text{initial}}| = 2 \text{ x } 10^{-3}$.

The stress amplitude has been calculated with different trigonometric equations and FE simulation shown in the sections 3.2.7 and 3.3.2.

The blue points represent bending tests, the red one torsion tests. It can be observed that the torsion lifetime is shorter than the bending one, and that the results are widely distributed. Indeed, for the same stress amplitude, the results are scattered along more than three orders of magnitude.

Finally, the filled markers correspond to a relative resonant frequency decrease of -2 x 10^{-3}. They represent the crack initiation lifetime (see 5.1.2), equivalent to the lifetime to failure in the VHCF regime. The unfilled ones have been stopped after an increase of 2 x 10^{-3} of the resonant frequency, and are marked by an arrow. They can be considered as run-outs as no crack initiation occurs.

The limit represented by the dotted line between the filled and the unfilled markers is named 'critical amplitude' (see section 5.3). It is equal to about 12 MPa for bending tests, and 6 MPa for torsion tests.

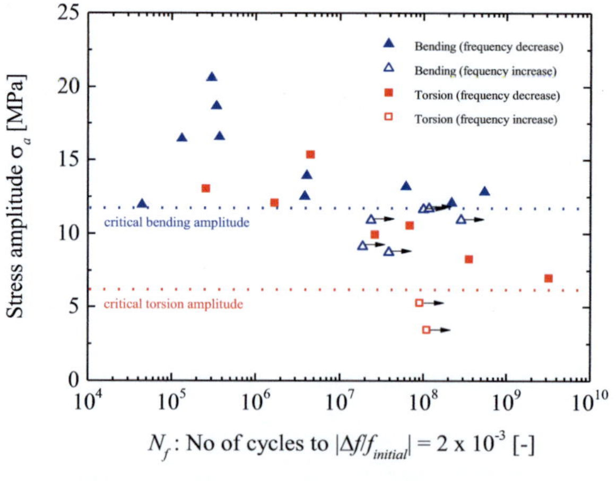

Fig. 5.16: Al lifetime diagram – S-N curve – bending tests in blue, torsion tests in red – filled markers represent a decrease of $\Delta f/f_{\text{initial}} = -2 \times 10^{-3}$ of the sample resonant frequency, unfilled ones an increase of 2 x 10^{-3} – the dotted lines are the critical stress amplitudes separating the decrease and increase phenomena.

5.3.2.2 Sample list

Tab. 8.4 (see Appendix) summarizes the successful Al micro-bending fatigue tests. It also lists the sample thickness after electropolishing. Finally the lifetime criterion is shown and corresponds to a certain $\Delta f / f_{\text{initial}}$ decrease of -2×10^{-3}. On the contrary, in case of an increase of $\Delta f / f_{\text{initial}} = 2 \times 10^{-3}$ the test was stopped.

The Al torsion tests are listed in **Tab. 8.5** (see Appendix). This table is similar to **Tab. 8.4**.

5.3.2.3 Al - Bending - Decrease of the resonant frequency

The following section presents typical damage on an Al sample after a bending test where the relative resonant frequency decreased of $\Delta f / f_{\text{initial}} = -2 \times 10^{-3}$.

Fig. 5.17 is an optical micrograph of Al23. The optically dark areas, as marked by the red circles, are post fatigue extrusions.

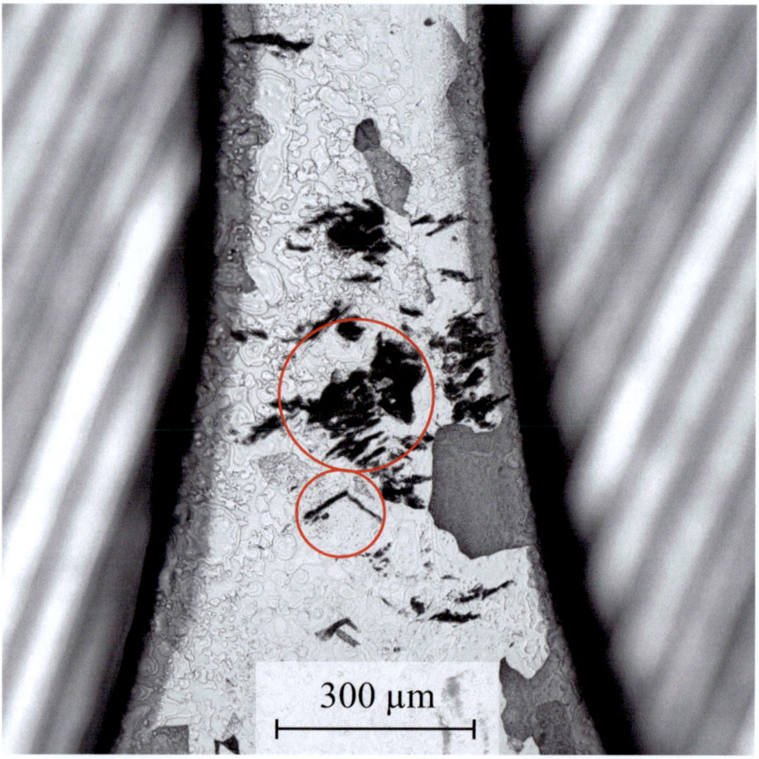

Fig. 5.17: Al23: Bending – decrease $\Delta f/f_{\text{initial}} = -2 \times 10^{-3}$ – overview – optical micrograph of the sample surface – post fatigue extrusions emphasized by red circles. The light gray marks are electropolishing artefacts and do not influence the fatigue mechanisms

Fig. 5.18 is a scanning electron micrograph of the two locations marked by red circles in **Fig. 5.17**. The lines or dark areas marked by white arrows are extrusions.

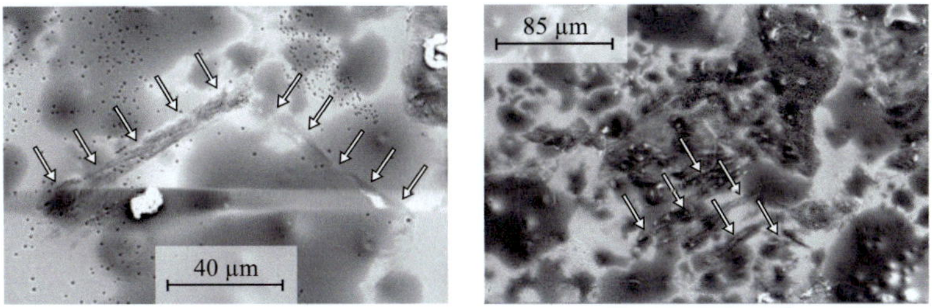

(a) (b)

Fig. 5.18: Al23: Bending – decrease $\Delta f/f_{initial} = -2 \times 10^{-3}$ – scanning electron micrographs of the sample surface– post fatigue extrusions marked by white arrows.

The same observations can be made for Al33. The biggest post fatigue extrusions are marked by red circles in **Fig. 5.19**.

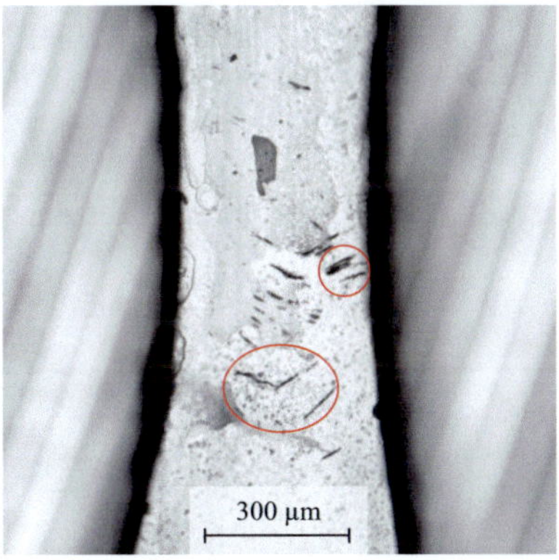

Fig. 5.19: Al33: Bending – decrease $\Delta f/f_{initial} = -2 \times 10^{-3}$– overview – optical micrograph of the sample surface – post fatigue extrusions emphasized by red circles. The light gray marks are electropolishing artefacts and do not influence the fatigue mechanisms

The optically dark lines marked by red circles in **Fig. 5.19** are extrusions, as can be seen along the white arrows in **Fig. 5.20**.

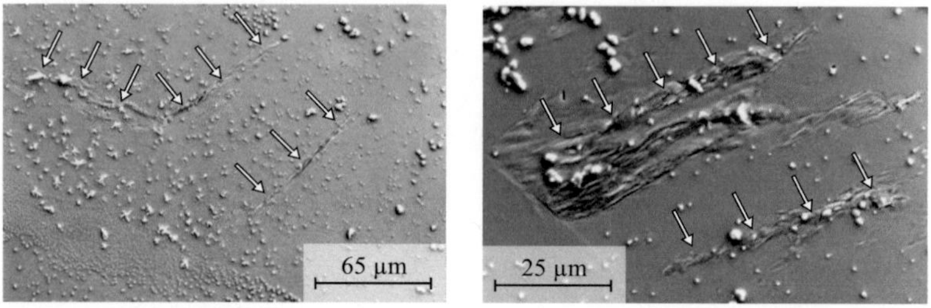

(a) (b)

Fig. 5.20: Al33: Bending – decrease $\Delta f/f_{\text{initial}} = -2 \times 10^{-3}$ – scanning electron micrographs of the sample surface – post fatigue extrusions marked by white arrows.

FIB cross sections are necessary to know if pores are situated under the extrusions, unfortunately this has not been done because of a matter of means. Nevertheless, the following section shows similar damage where FIB cross sections reveal subsurface pores (**Fig. 5.23**).

5.3.2.4 Al – Bending – Increase of the resonant frequency

The following sample, Al24, shows typical results after a test with a subcritical stress amplitude. An increase of the resonant frequency of $\Delta f/f_{\text{initial}} = 2 \times 10^{-3}$ has been measured.

Fig. 5.21 is a scanning electron micrograph where the surface of Al24 can be observed. The white circles point out some fatigued grains, where extrusions (white lines or points) can be seen.

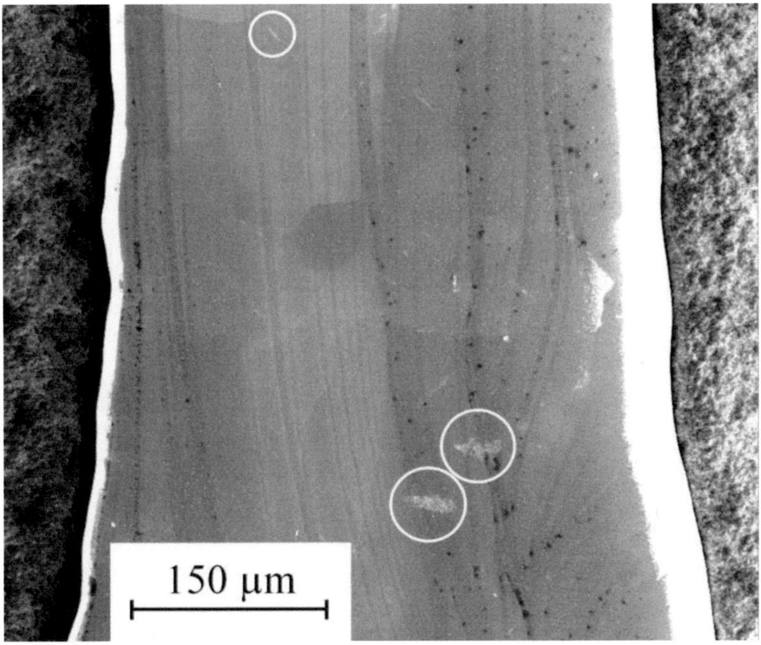

Fig. 5.21: Al24: Bending – increase $\Delta f/f_{initial} = 2 \times 10^{-3}$ – overview – scanning electron micrograph of the sample surface – post fatigue extrusions emphasized by white circles. The light gray vertical lines along the sample axis were there before fatigue and are electropolishing artefacts. It can be observed that those vertical lines do not interfere with the damage evolution, since extrusions grew in single grains despite them.

A higher magnification scanning electron micrograph has been taken at the grains marked by the white circles in **Fig. 5.21**. Few single extrusions along the grain boundary can be seen on the sample surface (e.g. **Fig. 5.22**a). Though, no subsurface damage can be observed at the FIB cross section (**Fig. 5.22**b). The white dashed lines in **Fig. 5.22**a represent the location of the FIB cross section.

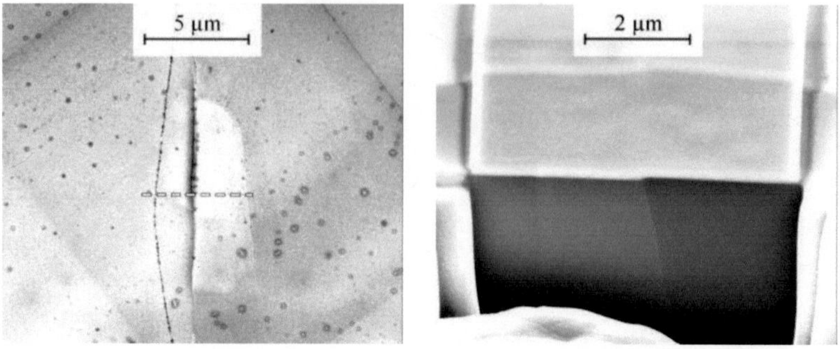

(a) (b)

Fig. 5.22: Al24: Bending – increase $\Delta f/f_{\text{initial}} = 2 \times 10^{-3}$ – scanning electron micrographs of the sample surface (a) and of a FIB cross section (b) – no subsurface damage can be observed under the extrusion.

Some single damaged grains (**Fig. 5.23**a) on Al24 show, however, subsurface damage. **Fig. 5.23**b shows some hundreds nm long pores, below the sample surface.

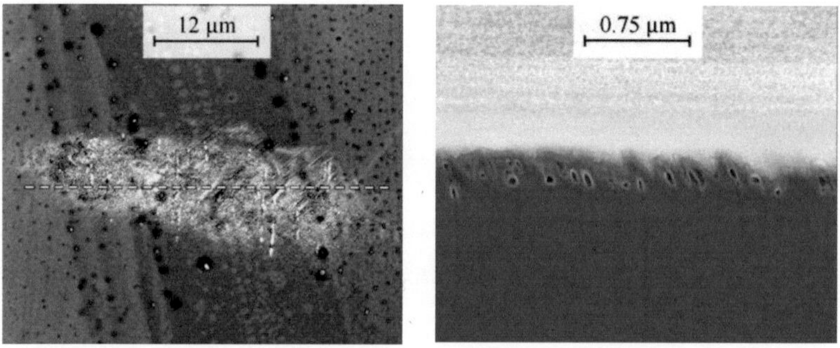

(a) (b)

Fig. 5.23: Al24: Bending – increase $\Delta f/f_{\text{initial}} = 2 \times 10^{-3}$ – scanning electron micrographs of the sample surface (a) and of a FIB cross section (b) – small subsurface pores can be observed under a fatigued grain covered by extrusions.

In some cases the pores, below the surface, are a few µm in length and can reach the sample surface (**Fig. 5.24**b).

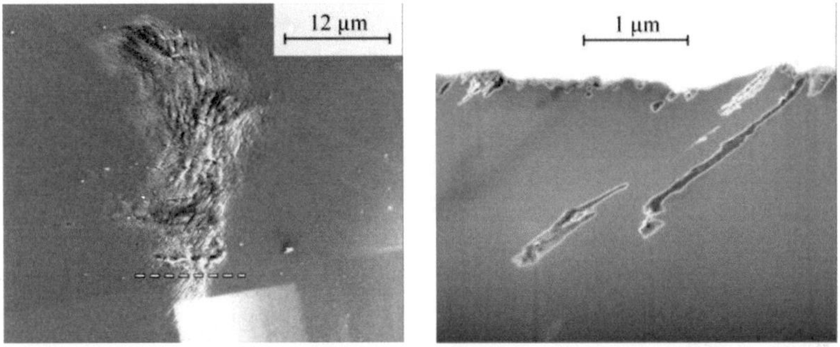

(a) (b)

Fig. 5.24: Al24: Bending – increase $\Delta f/f_{initial} = 2 \times 10^{-3}$ – scanning electron micrographs of the sample surface (a) and of a FIB cross section (b) – long subsurface pores can be observed under a fatigued grain covered by extrusions.

In conclusion, the major difference in terms of damage between the decrease and the increase of the resonant frequency is the quantity of fatigued grains on the sample surface. Indeed comparing **Fig. 5.17** and **Fig. 5.19** with **Fig. 5.21**, in case of an increase, there are fewer grains showing extrusions. One can also observed that they are every time constrained in a single grain and do not propagate to the neighbor. In both cases the damage is nevertheless similar.

5.3.2.5 Al - Torsion - Decrease of the resonant frequency

The next sample, Al10, is a typical example of a test where the stress amplitude was above the critical amplitude, so called overcritical, and where a decrease of the resonant frequency of $\Delta f/f_{\text{initial}} = -2 \times 10^{-3}$ has been measured.

Fig. 5.25 is a scanning electron micrograph of the sample surface. The dark elongated areas in the middle of the sample are fine post fatigue extrusions. The highest extrusions are emphasized with circles. **Fig. 5.26** is a higher magnification micrograph of these damaged grains.

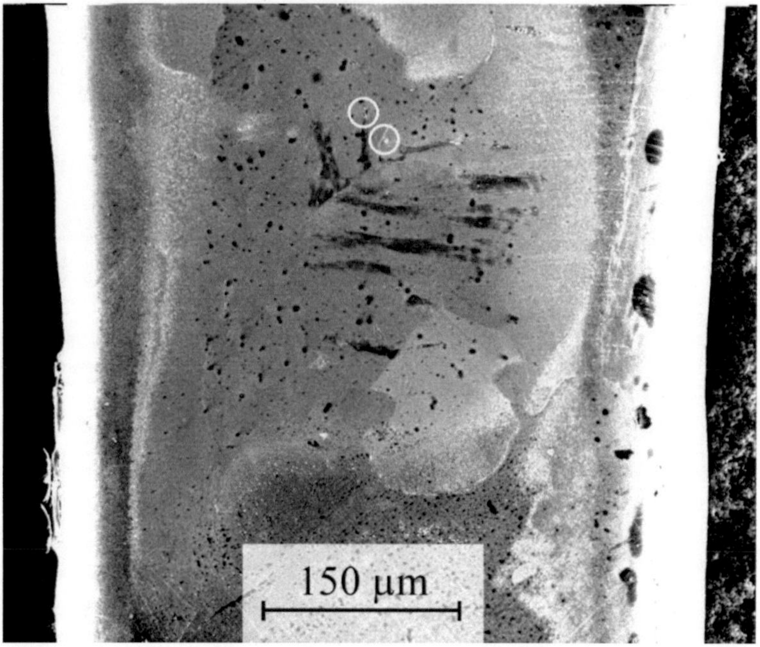

Fig. 5.25: Al10: Torsion – decrease $\Delta f/f_{initial} = -2 \times 10^{-3}$ – overview – scanning electron micrograph of the sample surface – post fatigue extrusions emphasized by white circles. Other than the post fatigue damage, some grains appear with different gray value and some dark dots cover the sample surface, this is due to the electropolishing process. These defects do not influence the fatigue damage evolution though.

The really dark and long areas are extrusions as it has also been seen in the previous paragraphs (e.g. **Fig. 5.17** and **Fig. 5.19**).

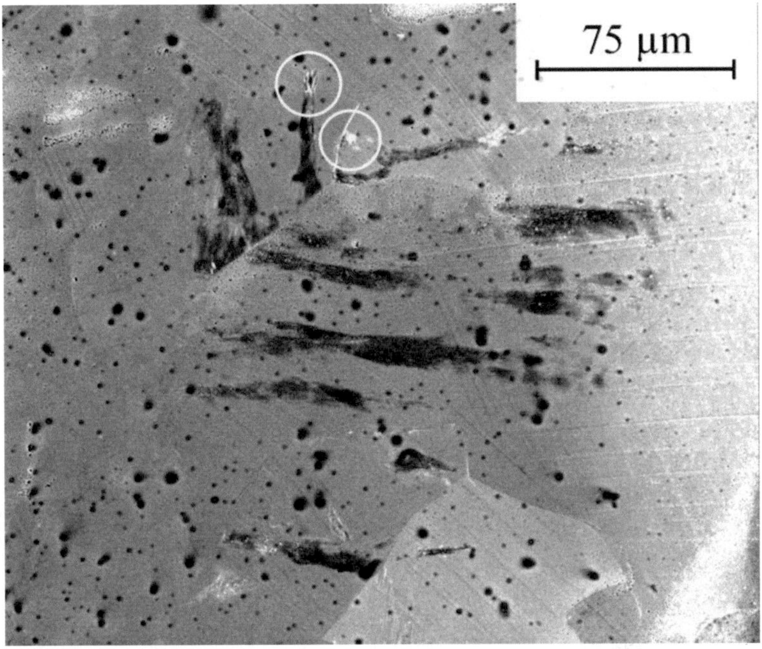

Fig. 5.26: Al10: Torsion – decrease $\Delta f/f_{initial} = -2 \times 10^{-3}$ – overview zoomed in – scanning electron micrograph of the sample surface – post fatigue extrusions emphasized by white circles.

Some extrusions (**Fig. 5.27**a representing the right white circle of **Fig. 5.26**) crossed the grain boundary and continued to propagate. They were starting to grow in another direction when the test has been stopped. **Fig. 5.27**b is a FIB cross section, some pores can be observed under the surface or at the bottom of intrusions.

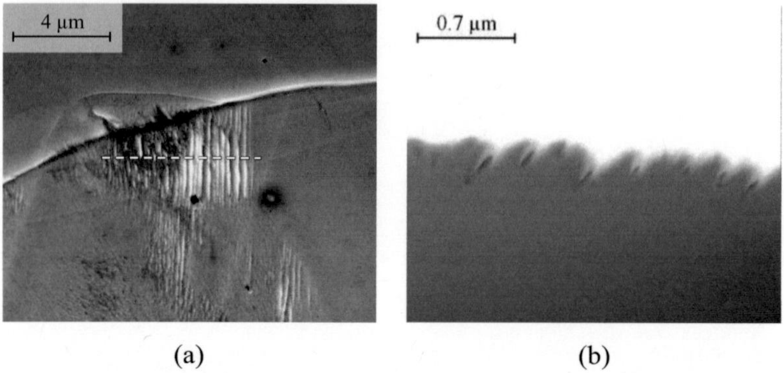

Fig. 5.27: Al10: Torsion – decrease $\Delta f/f_{\text{initial}} = -2 \times 10^{-3}$ – scanning electron micrographs of the sample surface (a) and of a FIB cross section (b) – small subsurface pores can be observed under a fatigued grain at the bottom of intrusions.

The same observations in another grain can be made. **Fig. 5.28** represents the left white circle of **Fig. 5.26**.

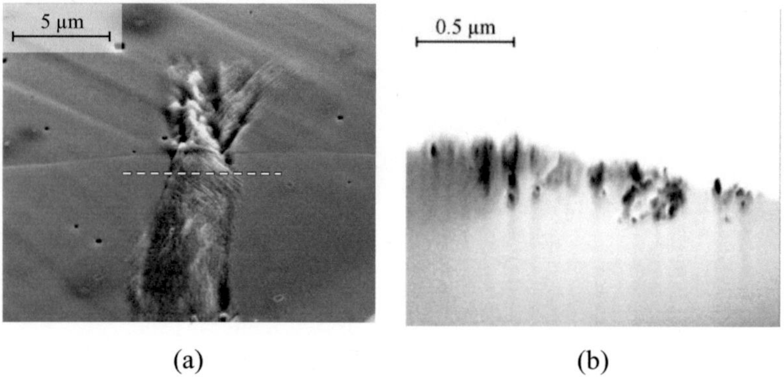

Fig. 5.28: Al10: Torsion – decrease $\Delta f/f_{\text{initial}} = -2 \times 10^{-3}$ – scanning electron micrographs of the sample surface (a) and of a FIB cross section (b) – small subsurface pores can be observed under a fatigued grain covered by extrusions/intrusions.

Al19 (**Fig. 5.29**) is another example of the decrease of the resonant frequency of $\Delta f / f_{\text{initial}} = -2 \times 10^{-3}$ after a torsion test.

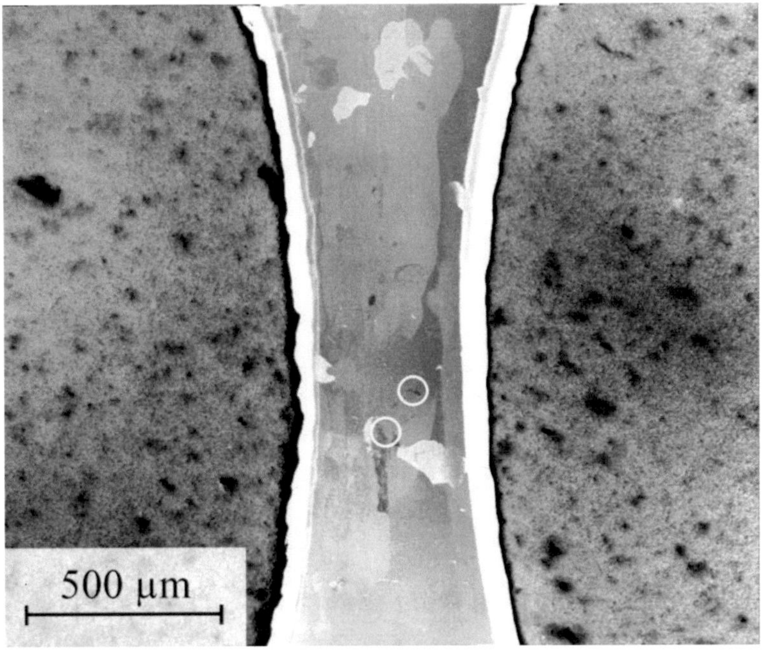

Fig. 5.29: Al19: Torsion – decrease $\Delta f/f_{\text{initial}} = -2 \times 10^{-3}$ – overview – scanning electron micrograph of the sample surface – post fatigue extrusions emphasized by white circles. This overview shows clearly the influence of the electropolishing process on the sample surface. Depending on the grain orientation, the process reacts differently with the material, and grains appear with a different gray value in the SEM.

The two white circles point out the highest post fatigue extrusions.

Fig. 5.30 shows two higher magnification micrographs of the previous presented damaged grains.

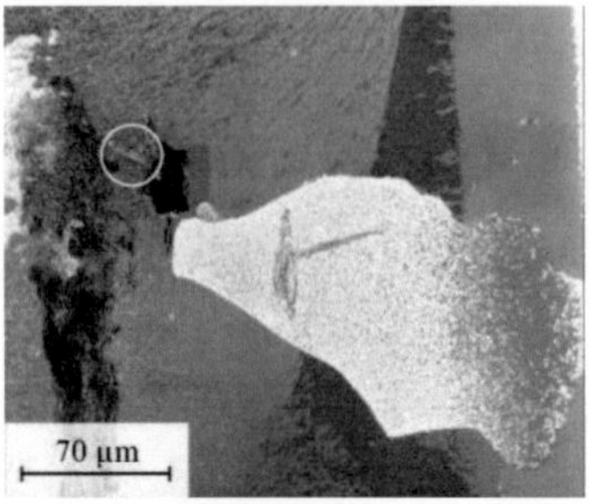

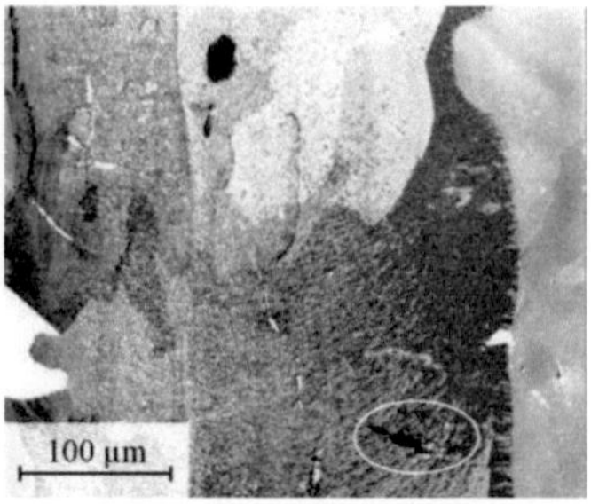

(a) (b)

Fig. 5.30: Al19: Torsion – decrease $\Delta f/f_{initial} = -2 \times 10^{-3}$ – overview zoomed in – scanning electron micrographs of the sample surface – post fatigue extrusions emphasized by white circles.

A couple of single extrusions can be found on the sample surface. **Fig. 5.31**a corresponds to a higher magnification of **Fig. 5.30**b, and shows a large intragranular extrusion. **Fig. 5.31**b shows a FIB cross section (at the location of the dashed line in **Fig. 5.31**a) where only one pore (white arrow) at the bottom of an intrusion can be seen.

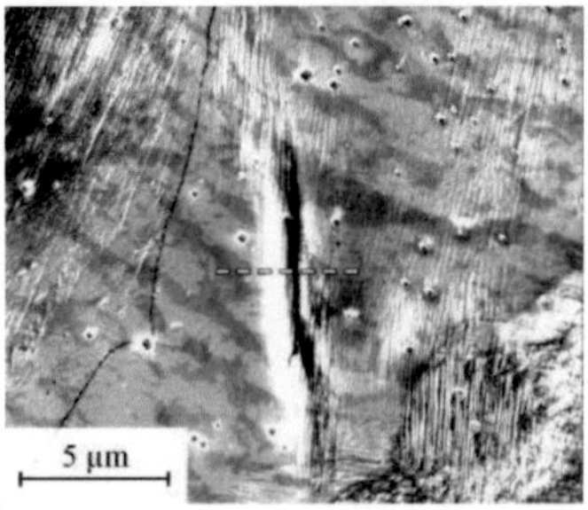

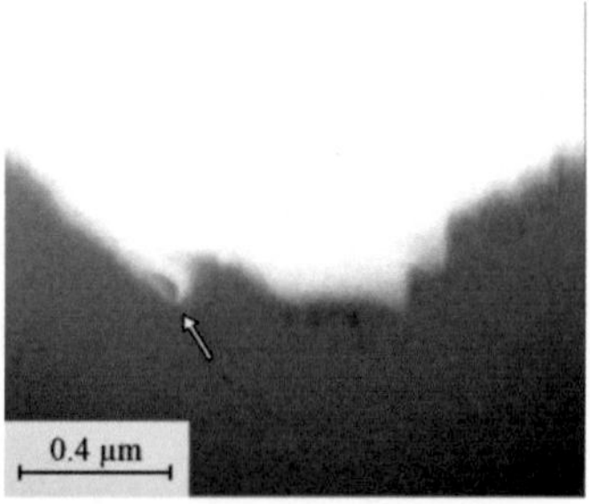

(a) (b)

Fig. 5.31: Al19: Torsion – decrease $\Delta f/f_{initial} = -2 \times 10^{-3}$in – scanning electron micrographs of the sample surface (a) and of a FIB cross section (b) – only one subsurface pore (see white arrow).

Some single damaged grains can also be found on the sample surface. **Fig. 5.32** is an example of one of those, and is a higher magnification of **Fig. 5.30**b. Extrusions were formed during the test, and are constrained in the grain.

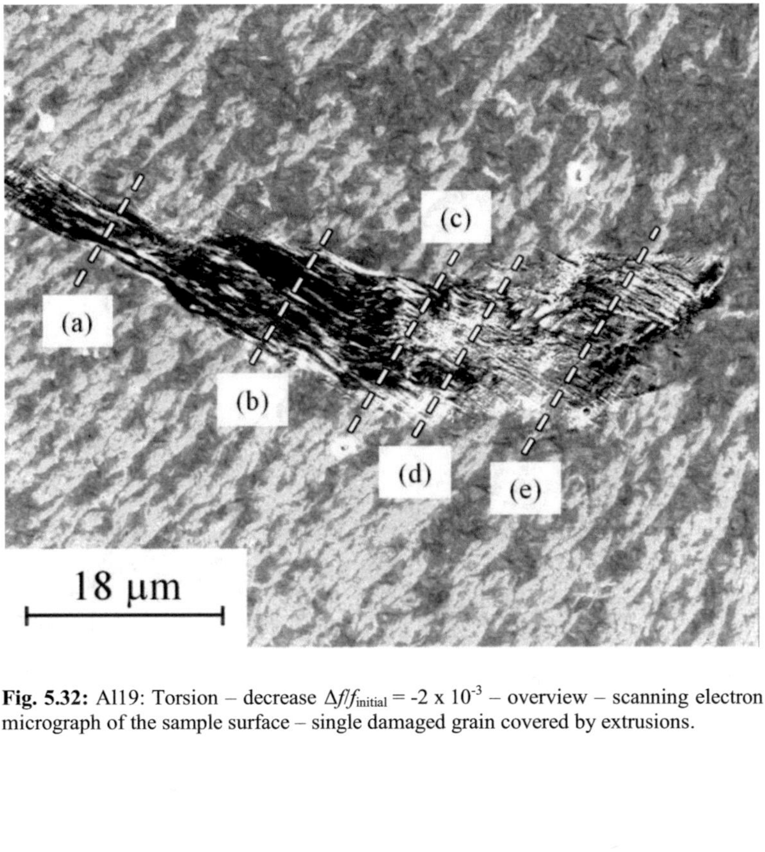

Fig. 5.32: Al19: Torsion – decrease $\Delta f/f_{initial} = -2 \times 10^{-3}$ – overview – scanning electron micrograph of the sample surface – single damaged grain covered by extrusions.

Fig. 5.33 shows different FIB cross sections along the grain, where the pore evolution beneath the surface of the damaged grain can be observed. The different dashed lines (a, b, c, d, and e) in **Fig. 5.32** represent the different FIB cut locations detailed in **Fig. 5.33**.

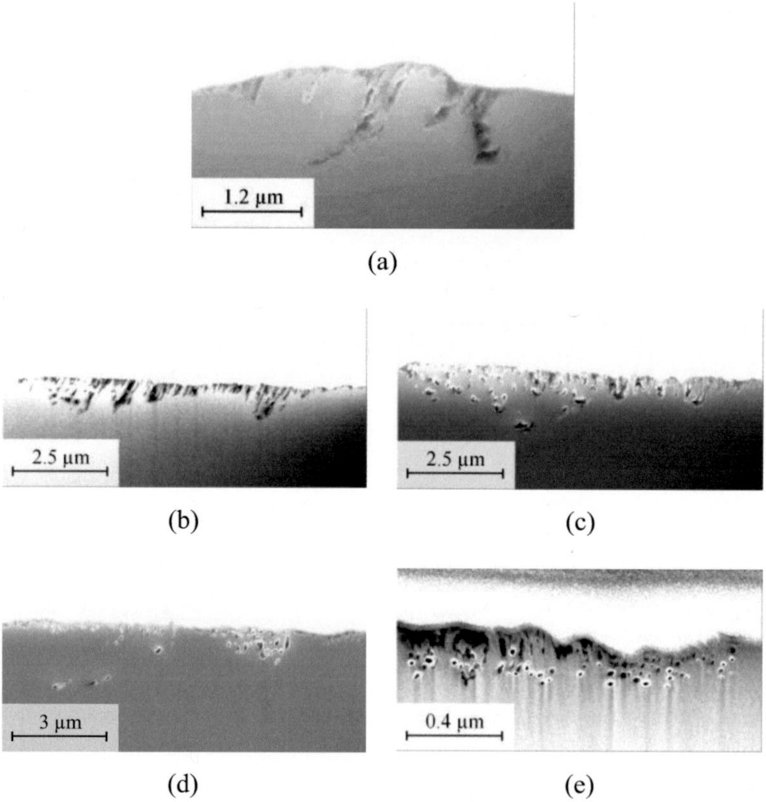

Fig. 5.33: Al19: Torsion – decrease $\Delta f/f_{initial} = -2 \times 10^{-3}$ – scanning electron micrographs of FIB cross sections along a single damaged grain – the subsurface pore evolution can be observed.

5.3.2.6 Al – Torsion – Increase of the resonant frequency

The following section shows the damage on an Al sample after a torsion test under a subcritical stress amplitude where the resonant frequency increased to $\Delta f/f_{\text{initial}} = 2 \text{ x } 10^{-3}$.

Fig. 5.34 is an overview of the sample surface. Only few post fatigue damage sites can be observed. The white circles point them out.

Fig. 5.34: Al14: Torsion – increase $\Delta f/f_{\text{initial}} = 2 \text{ x } 10^{-3}$ – overview – scanning electron micrograph of the sample surface – post fatigue extrusions emphasized by white circles. The long thin dark lines on the sample surface were already there before fatigue and are probably related to the sample fabrication.

Fig. 5.35 is an example of one of the few damaged grains; it corresponds to the bottom white circle in the previous figure. Some extrusions can be observed, but only few pores appear beneath the surface.

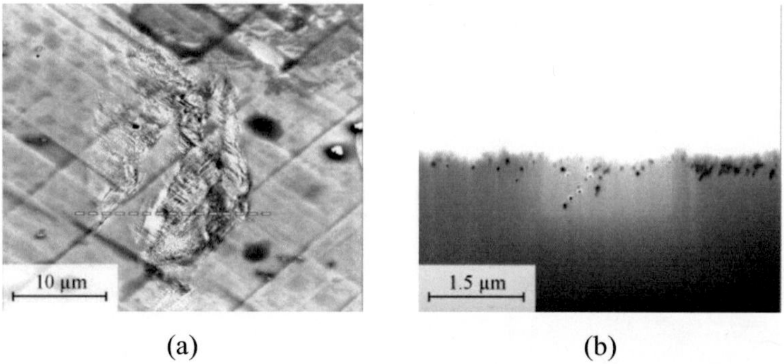

Fig. 5.35: Al14: Torsion – increase $\Delta f/f_{initial} = 2 \times 10^{-3}$ – scanning electron micrographs of the sample surface (a) and of a FIB cross section (b) – small subsurface pores can be observed in a fatigued grain.

Fig. 5.36a is a higher magnification of the top white circle in **Fig. 5.34**. A crack initiated between two grains, but stays on the sample surface, no subsurface damage is visible in **Fig. 5.36**b.

Fig. 5.36: Al14: Torsion – increase $\Delta f/f_{initial} = 2 \times 10^{-3}$ – scanning electron micrographs of the sample surface (a) and of a FIB cross section (b) – no subsurface damage can be observed.

The same observations as in section 5.3.2.4 are made. A detailed microscopic investigation of the different fatigued samples shows in case of an increase of the resonant frequency that the extrusions and damaged grains are isolated and less in number.

5.3.3 OFHC copper fatigue tests

5.3.3.1 Lifetime until $|\Delta f/f_{initial}| = 2 \times 10^{-3}$

As already described in 5.3.2.1 for the Al tests, **Fig. 5.37** shows the Cu test results, with the lifetime criterion explained in 5.1.2. $|\Delta f / f_{\text{initial}}| = 2 \text{ x } 10^{-3}$.

The blue points represent bending tests, the red one torsion tests and the magenta one the only multiaxial test. It can be observed that the results are widely distributed. Indeed on the same stress horizon, the results are scattered by about three orders of magnitude.

Finally, the filled markers correspond to a relative resonant frequency decrease of -2×10^{-3}. The unfilled ones have been stopped after an increase of 2×10^{-3}. The limit represented by the dotted line between the filled and the unfilled markers is the 'critical amplitude' (see section 5.3). It is equal to about 70 MPa for the bending tests.

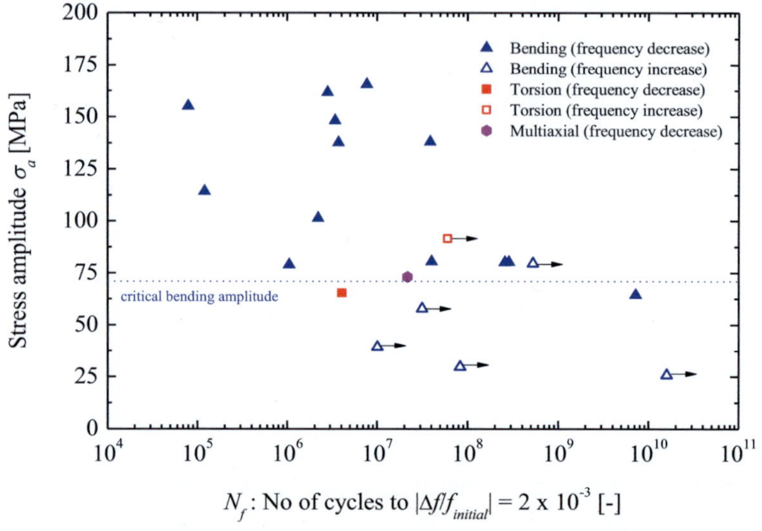

Fig. 5.37: Cu lifetime diagram – S-N curve – bending test in blue, torsion test in red and multiaxial test in magenta – filled markers represent a decrease of $\Delta f/f_{initial} = -2 \times 10^{-3}$ of the sample resonant frequency, unfilled ones an increase of 2×10^{-3} – the dotted line represents the critical stress amplitude separating the decrease and increase phenomena.

5.3.3.2 Sample list

Tab. 8.6 (see Appendix) summarizes the successful Cu micro-bending fatigue tests. It also lists the sample thickness after electropolishing. Finally the lifetime criterion is shown and corresponds to a certain $\Delta f / f_{\text{initial}}$ value. In case of a decrease of the resonant frequency $\Delta f / f_{\text{initial}}$ is equal to -2×10^{-3} on the contrary in case of an increase of 2×10^{-3}, the test was stopped.

In the same way, **Tab. 8.7** shows the torsion tests, and finally **Tab. 8.8** lists the single multiaxial test.

5.3.3.3 Experiment at the critical stress amplitude

The test with Cu34 reached the VHCF regime ($N_f = 7.2023 \times 10^9$) under a bending stress amplitude of about $\sigma_a = 64$ MPa. It has to be noted that this value is close to the critical stress amplitude.

The sample behavior reveals a high interest of what can happen in the VHCF regime. Indeed even if at the end of the test the relative resonant frequency decreased of $\Delta f / f_{\text{initial}} = -2 \times 10^{-3}$, it increased at the beginning of the test, and was equal to $\Delta f / f_{\text{initial}} = 1.78 \times 10^{-3}$ after $N_f = 3.99 \times 10^8$. This is the only test where such a behavior has been observed.

Tab. 8.9 shows the relative resonant frequency change for different number of cycles along the whole test. The frequency first linearly increased (such as the upper curve in **Fig. 5.15**) before starting to decrease until reaching the stop criterion. **Fig. 5.38** shows the decrease of the relative resonant frequency. The slope seems to be linear from $N_f = 3.99 \times 10^8$ to about $N_f = 5.76 \times 10^9$, and starts to decrease faster.

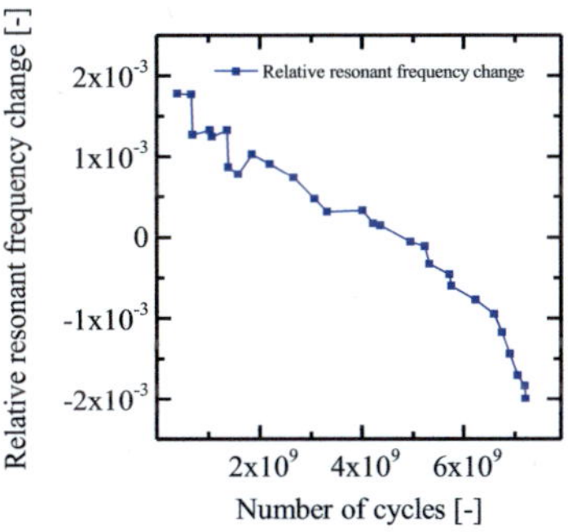

Fig. 5.38: Cu34 – The relative resonant frequency change is given as a function of the number of cycles. The frequency first increased until $N_f = 3.99 \times 10^8$, then started to decrease and stopped at $\Delta f / f_{\text{initial}} = -2 \times 10^{-3}$.

5.3.3.4 Cu – Bending – Decrease of the resonant frequency

The following section presents typical damage on a Cu-sample after a bending test with an overcritical stress where the resonant frequency decreased of $\Delta f / f_{\text{initial}} = -2 \times 10^{-3}$. The damage in Cu differs from the one in Al, instead of having subsurface pores below some extrusions, micro-cracks are formed at the bottom of intrusions.

Fig. 5.39 is a scanning electron micrograph overview of the sample surface of Cu13. Few grains are damaged after fatigue and marked by white circles.

Looking closer to the damaged grain marked by the top white circle in **Fig. 5.39**, few extrusion/intrusion pairs can be observed (**Fig. 5.40**a). **Fig. 5.40**b is a scanning electron micrograph of a FIB cross section and shows 500 nm to 1 µm deep micro-cracks (pointed out by white arrows), growing from the sample surface at the bottom of intrusions.

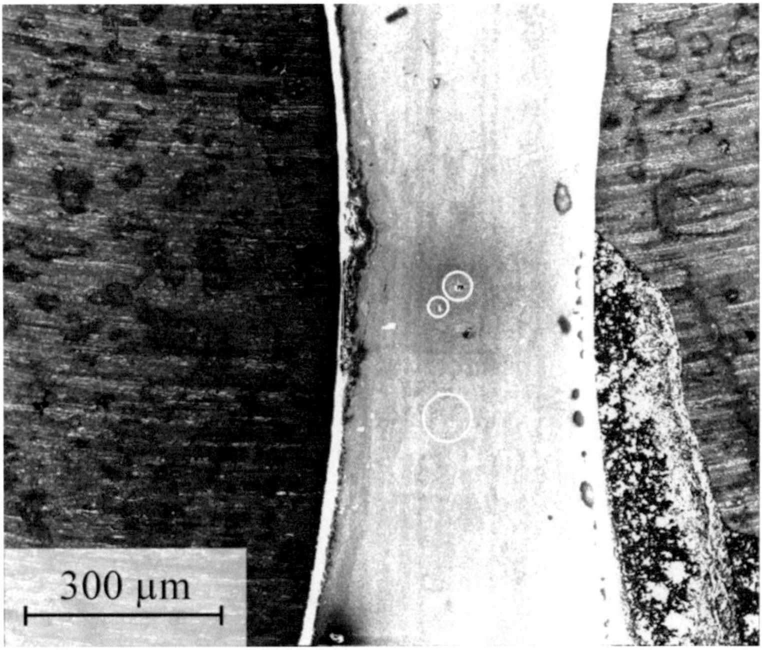

Fig. 5.39: Cu13: Bending – decrease $\Delta f/f_{initial} = -2 \times 10^{-3}$ – overview – scanning electron micrograph of the sample surface – post fatigue extrusions emphasized by white circles. Some light gray vertical lines can be observed all along the sample axis, those are rest of the rolling fabrication process. This sample was fixed on an aluminum sample holder with some conductive silver paint; the drops on the right side of the sample and the dark area on the left side are rests of this paint on the sample surface. These rests have been involuntary added to the surface during the sample preparation for SEM and not before fatigue. This did not influence the fatigue processes.

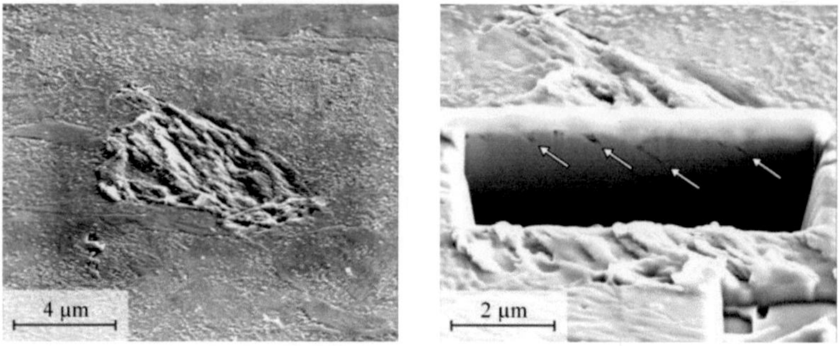

(a) (b)

Fig. 5.40: Cu13: Bending – decrease $\Delta f/f_{initial} = -2 \times 10^{-3}$ – scanning electron micrographs of the sample surface (a) and of a FIB cross section (b) – subsurface micro-cracks can be observed in a fatigued grain at the bottom of intrusions.

In **Fig. 5.41**, the same observations as in **Fig. 5.40** can be made for the grain marked by the middle white circle in **Fig. 5.39**. Nevertheless **Fig. 5.41**c is an ion-enhanced scanning electron micrograph and gives complementary information on the grain morphology. It can be seen that the extrusions are completely constrained in this only one grain.

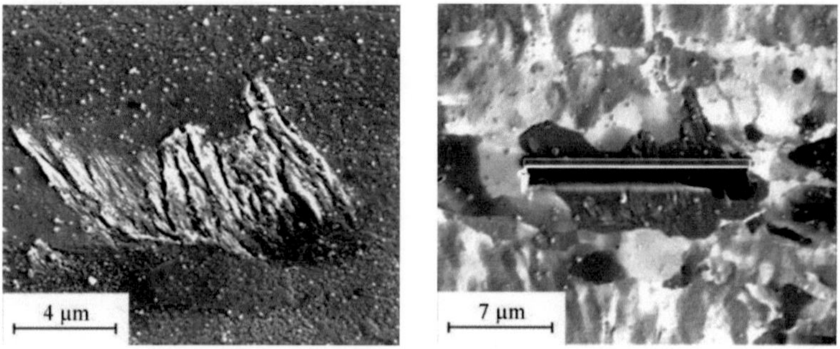

(a) (b)

Fig. 5.41: Cu13: Bending – decrease $\Delta f/f_{initial} = -2 \times 10^{-3}$ – scanning electron micrographs of the sample surface (a) and of an ion-enhanced FIB cross section (b) – subsurface micro-cracks can be observed at the bottom of intrusions as shown with an higher magnification in **Fig. 5.3**.

On the same sample, some other damage (all marked by the bottom white circle in **Fig. 5.39**) look like the very first crack initiation (**Fig. 5.42**a, **Fig. 5.43**a, **Fig. 5.44**a). It differs from the extrusion pattern seen before. Several 1 μm long single micro-cracks can be observed.

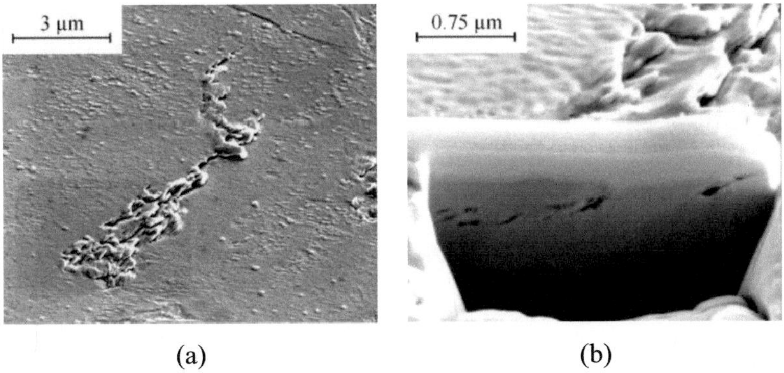

Fig. 5.42: Cu13: Bending – decrease $\Delta f/f_{initial} = -2 \times 10^{-3}$ – scanning electron micrographs of the sample surface (a) and of a FIB cross section (b) – subsurface intragranular micro-cracks.

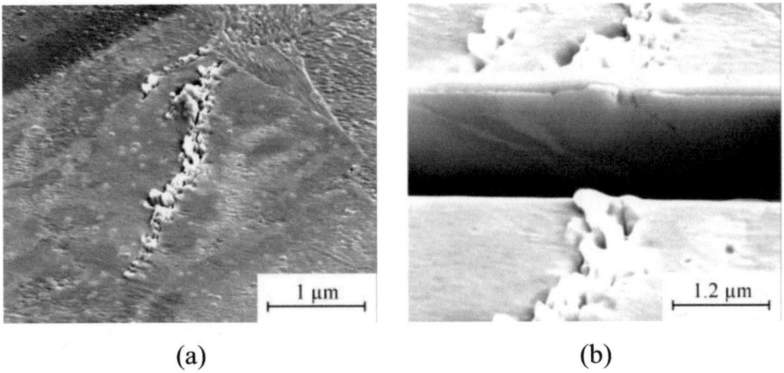

Fig. 5.43: Cu13: Bending – decrease $\Delta f/f_{initial} = -2 \times 10^{-3}$ – scanning electron micrographs of the sample surface (a) and of a FIB cross section – subsurface intragranular micro-crack.

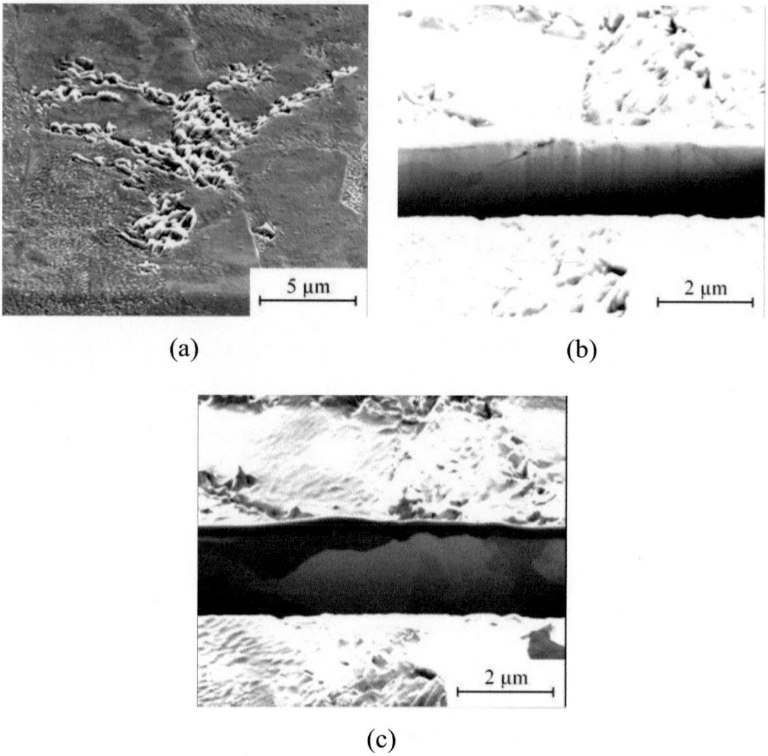

Fig. 5.44: Cu13: Bending – decrease $\Delta f/f_{initial} = -2 \times 10^{-3}$ – scanning electron micrographs of the sample surface (a), of a FIB cross section (b) and of an ion-enhanced FIB cross section (c) – subsurface intragranular micro-crack.

The ion-enhanced micrographs (**Fig. 5.42**b, **Fig. 5.43**b, **Fig. 5.44**b) show different contrasts, due to grains with different crystallographic microstructures. The micro-cracks are observed within single grains and are thereby intragranular.

5.3.3.5 Cu – Bending – Increase of the resonant frequency

This section presents the only one sample (**Fig. 5.45**) which has been analyzed within an SEM after a bending test with a subcritical stress where the resonant frequency increased to $\Delta f / f_{\text{initial}} = 2 \times 10^{-3}$. Just few damage sites can be observed on the surface.

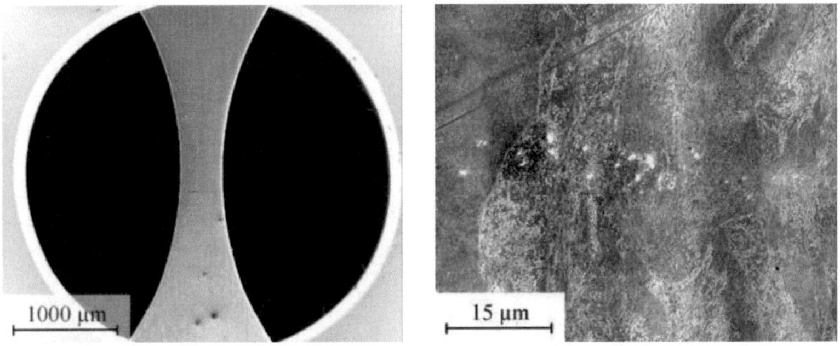

(a) (b)

Fig. 5.45: Cu15: Bending – increase $\Delta f/f_{\text{initial}} = 2 \times 10^{-3}$ – scanning electron micrographs of the sample surface overview (a) and of a rare surface damage (b).

5.3.3.6 Cu – Multiaxial loading – Decrease of the resonant frequency

Finally, in case of a multiaxial loading, after the resonant frequency decreased to $\Delta f / f_{\text{initial}} = -2 \times 10^{-3}$, damage on the surface or even in the sample was developed following two orientations.

Fig. 5.46 is a scanning electron micrograph of Cu2. White circles emphasize extrusion/intrusion pattern, on which is focused in the three next figures.

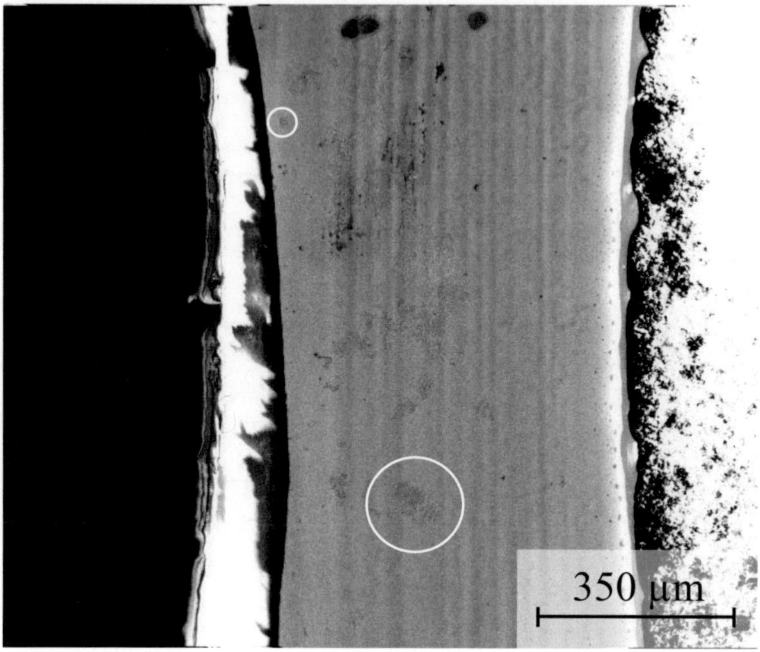

Fig. 5.46: Cu2: Multi – decrease Δf/finitial = -2 x 10-3 – overview – scanning electron micrograph of the sample surface – post fatigue extrusions emphasized by white circles. The vertical gray lines are the rolling direction, and the bright and dark area on the left of the sample is some post fatigue conductive silver paint. These did not influence the fatigue processes.

The fatigue damage follows two directions, as can be observed in **Fig. 5.47**. Some extrusions in **Fig. 5.47**a cross each other and the micro-cracks have different orientations (**Fig. 5.47**b, **Fig. 5.49**b). These different orientations have been indicated with white arrows.

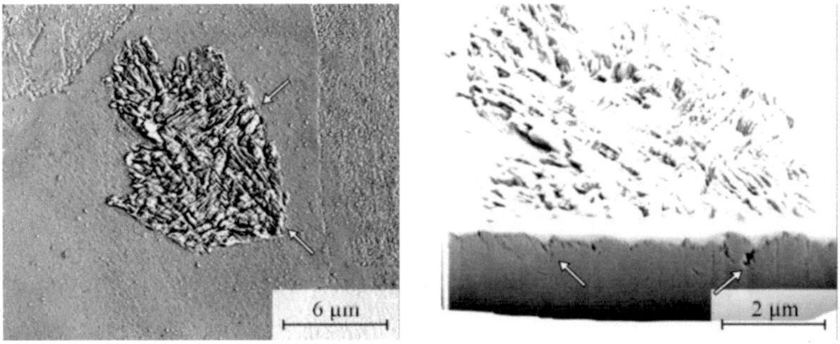

(a) (b)

Fig. 5.47: Cu2: Multi – decrease $\Delta f/f_{initial} = -2 \times 10^{-3}$ – scanning electron micrographs of the sample surface (a) and of a FIB cross section (b) – subsurface micro-cracks – two orientations indicated with white arrows.

The same extrusions following two orientations are observed in **Fig. 5.48** and **Fig. 5.49**.

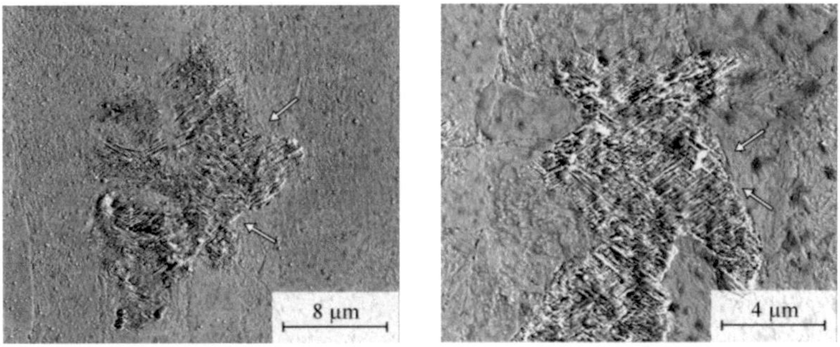

(a) (b)

Fig. 5.48: Cu2: Multi – decrease $\Delta f/f_{initial} = -2 \times 10^{-3}$ – scanning electron micrographs of the sample surface (a) and (b) – surface damage following two orientations indicated with white arrows.

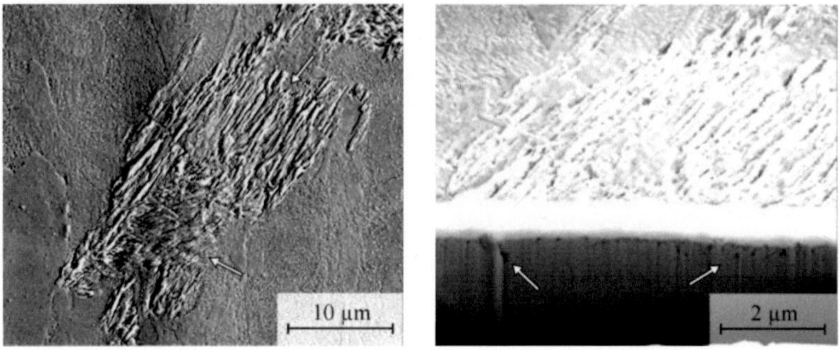

(a) (b)

Fig. 5.49: Cu2: Multi – decrease $\Delta f/f_{\text{initial}} = -2 \times 10^{-3}$ – scanning electron micrograph of the sample surface (a) and of a FIB cross section (b) – subsurface micro-cracks – two orientations indicated with white arrows.

Even if the extrusions follow two orientations, the same observations are made, subsurface micro-cracks are found at intrusion bottom.

6 Discussion

6.1 Damage and crack initiation detection methods

The term crack initiation should first be clarified since it is directly correlated to the different damage and crack observation methods. Zimmermann [60] concluded in her review that the nucleation of a fatigue crack could be described as the damage evolution from the atomic scale up to 1 µm. It is followed by micro-cracks and short cracks, with a size between 1 µm and 1 mm, and with everything above being considered as a long crack.

6.1.1 Existing damage detection methods

Compared to other techniques (e.g. replica technique, resistive measurements, thermography, Fourier analysis and their combination, (see detailed review in section 2.7), the new custom-built resonant fatigue setup generates additional insights on crack initiation. This new method can be considered as an in-situ damage tracking. Indeed, as soon as a relevant resonant frequency change is observed, the test can be stopped, and the damage evolution can be observed (as seen in section 5.2.1). This is for example not the case for the replica technique which only allows for the observation of the crack growth step by step when the test is arbitrarily stopped, without knowing if a particular damage evolution happened.

Other experimental setups have recently been developed to look at the early damage initiation; for example, in medium carbon steel considering generator power and specimen temperature [61] or in carbon fiber reinforced plastics using fast Fourier, short time Fourier, and Hilbert Huang transformation [62].

Nevertheless, none of them can observe a relative resonant frequency change corresponding to a crack length smaller than 1 µm.

In this work, cracks smaller than 500 nm have been observed in a Cu sample (see **Fig. 5.3**).

In contrast, smaller samples allow for a higher signal to noise ratio concerning the crack initiation, as the initiated cracks contribute much more to the sample behavior than in a macroscopic sample.

6.1.2 Custom-built resonant setup detection method

The implementation of the presented custom-built resonant setup allows for investigations of fatigue at a variety of size scales and loading scenarios.

The strong correlation between the resonant frequency change and the damage evolution (see 5.2.1) indicates that the increasing plasticity, resulting in extrusions, intrusions, crack initiation, and crack growth, can lead to a stiffness decrease of the sample and therefore directly impacts the resonant frequency. Since the experiments are strain controlled, the impact of plasticity on resonant frequency can be explained by a cyclic softening. This softening leads to a decrease in the stress amplitude and hence the slope of the stress-strain hysteresis is reduced. This decrease could be compensated, as it is in the work of Yang et al., by changing the position of the piezo actuator to maintain mean load [63, 64]. Nevertheless this is not the case in this work.

The method developed in this thesis has a very fine measurement resolution of the relative resonant frequency change of $\Delta f / f_{\text{initial}} = 5 \times 10^{-7}$ (as described in 5.1). This allows the accurate tracking of fatigue damage formation, especially prior to crack initiation, such as plasticity induced surface roughening and slip band formation (**Fig. 5.2**) at a state where it appears in only a few grains.

The orientation of these grains can be found performing Electron Backscatter Diffraction imaging (**Fig. 5.14**), providing information about the crystal lattice planes where damage accumulation initiates and later leads to fatigue failure. This setup is a novel way of experimentally examining damage evolution and crack initiation processes. Crack initiation and the subsequent short crack formation and growth play a particularly important role in the HCF and VHCF regimes, they very strongly determine the material lifetime.

The test in section 5.1 described the damage evolution in Ni (**Fig. 5.1**) with a maximal stress amplitude of $\sigma_a = 200$ MPa. This corresponds to what Mughrabi [41] described and follows three phases.

In the first phase (see section 5.1.1), the decrease of the resonant frequency is equal to $\Delta f / f_{\text{initial}} = -2 \times 10^{-4}$. It corresponds to the formation of the first slip bands in single grains (**Fig. 5.2**). Microstructural FIB cross sections show that there are still no cracks in this phase. In some rare cases, slip bands start to grow in adjacent grains.

In the second phase (see section 5.1.2), the resonant frequency decrease is equal to $\Delta f / f_{\text{initial}} = -2 \times 10^{-3}$. FIB cross sections show that crack nucleation appears in single grains on the sample surface (**Fig. 5.3**b). The very short cracks are formed exclusively on or in the existing slip bands (from the previous phase), see **Fig. 5.4**e. At the point where the crack meets the grain boundary, some defects develop in the neighboring grain also in the form of slip bands. Depending on the neighboring grain orientation, these slip bands stay either close to the grain boundary or can pass through the whole grain.

In order to understand the mechanisms of crack nucleation for this kind of micro-samples at this stage of the fatigue, optical micrographs and scanning electron micrographs have been regularly taken. The micrographs revealed evidence of slip band formation, as well as surface oxide thickening (**Fig. 5.6**) and micro-crack nucleation from surface oxides at PSBs (**Fig. 5.4**e).

Baumert et al. [19] also observed that oxide cracking leads to a drop of the resonant frequency (**Fig. 2.10**). FEM simulations of microcracks, physically consistent with their SEM and TEM observations, show a decrease of the resonant frequency corresponding to the one observed in their experiments. For a 5 x 0.5 μm² crack in a sample with a 20 x 11 μm² cross section, they observed a decrease close to

20 Hz when the micro-resonator has a resonant frequency of ~8 kHz. This corresponds to a maximal relative resonant frequency decrease of $\Delta f / f_{\text{initial}} = -2.5 \times 10^{-3}$, and validates the observation on the resonant frequency decrease of this work.

In the third phase of the damage evolution (see section 5.1.3), the resonant frequency decrease is equal to $\Delta f / f_{\text{initial}} = -2 \times 10^{-2}$. Some of the micro-cracks grow across several grains (**Fig. 5.5**). By means of EBSD (**Fig. 5.14**), it can be observed that grains where cracks propagate have a similar crystal orientation. A quantitative EBSD analysis of the grain orientation where the crack can or cannot grow is work in progress. This will be an important input for the DDD simulations of multi grain aggregates.

This third phase can also be described by the FE simulations from section 3.4.5. It is important to notify that the average nodal solution with the full graphics settings was selected in ANSYS. This guarantees that any kind of material, real constant, and geometric discontinuities were possible at the crack tip. Multiple cracks have been implemented in the sample with ANSYS to investigate their influence on the resonant frequency (see section 3.4). With a single 300 µm long und 30 µm deep crack, the bending resonant frequency sees a decrease of $\Delta f/f_{\text{initial}} = - 8.4 \times 10^{-4}$. For the worst case scenario, the maximal relative resonant frequency decrease is equal to $\Delta f/f_{\text{initial}} = - 1.77 \times 10^{-3}$ (**Fig. 3.33**). It corresponds to the resonant frequency decrease of the model with ‘two opposite cracks in one area’. This model has only two cracks implemented but sees a resonant frequency decrease almost as large as a model with four cracks. The two cracks are nevertheless positioned in the same transversal section and have a higher impact on the sample stiffness.

The maximal relative resonant frequency decrease of $\Delta f/f_{\text{initial}} = - 1.77 \times 10^{-3}$ observed in the simulation is close to what has been observed in the experiments, nevertheless only when the crack initiation takes place for $\Delta f/f_{\text{initial}} = - 2 \times 10^{-3}$. One can argue that the high number of micro-cracks observed in some grains in the experiments may be equivalent to the two long cracks in the simulation.

With the initiation of cracks, the stiffness of the sample is even further decreased. While the elastic effect of micro-cracks in a large sample cannot be picked up in the mechanical response due to the intrinsic and extrinsic damping, micro-cracks in a small sample volume will have a much higher signal to noise ratio and can be tracked. Here scaling effects are beneficial as the intrinsic damping scales with volume, and therefore is reduced with sample size in a cubic fashion while the crack initiation will only take place at the surface which scales in a quadratic fashion. With increasing number and length of cracks, the resultant impact on stiffness becomes more and more apparent. This may explain what seems to be a quadratic evolution of the resonant frequencies in **Fig. 3.33**.

Finally, although the setup is best suited for the detection of surface damage, it is also capable of detecting damage in the interior volume by tracking the frequency change. This internal damage onset would then be relatively difficult to locate (FIB and SEM required), when compared to surface damage, but it is possible.

6.2 Early damage evolution

6.2.1 Analysis of the first fatigue damage

Different damage growing rates can be interpreted from what has been observed in section 5.2, where damage evolution and relative resonant frequency change have been correlated.

Fig. 6.1 exactly illustrates how sensitive the damage evolution can be tracked (see **Tab. 8.1**, **Tab. 8.2** and **Tab. 8.3**) and observed (see **Fig. 5.8** and **Fig. 5.12**).

Nevertheless, the validity of all gray level ranges from **Fig. 5.11** and **Fig. 5.13** should be discussed. Looking at **Fig. 5.12**e, f, and g, it can be seen that the grain is almost entirely damaged. Therefore, the relative black pixel amount should be relatively high. This observation refutes the validity of the blue curve corresponding to the 0 to 50 gray level range. With a similar argument, looking at the little damage in

Fig. 5.12a, b, c, the validity of the 0 to 150 gray level range (black curve) may be questioned. For these reasons, only the results from the 0 to 100 gray level range (red curve) are discussed.

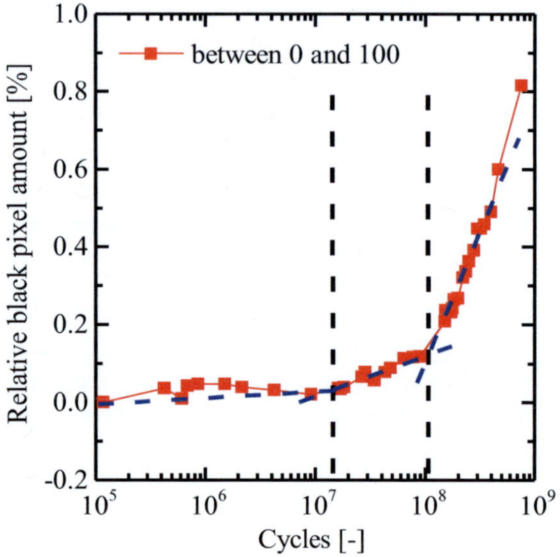

Fig. 6.1: Relative pixel amount between the gray levels 0 and 100 as function of the number of cycles – the blue dashed lines are asymptotes, revealing the slope of the curve – the vertical dark dashed lines separate regions where the relative black pixel amount rate along the number of cycles is different.

In **Fig. 6.1**, the gray level changes within the noise level of the measurement, but qualitatively (see **Fig. 5.11**a to i), increases until 1.5 x 10^7 cycles. Then, it increases slightly faster between 1.5 x 10^7 cycles and about 1 x 10^8 cycles.

On a logarithmic scale it finally increases faster after 10^8 cycles. The first part corresponds to the tests where extrusions only grow in single grains. The second part, where the growing rate is slightly higher, corresponds to the extrusion propagation in the neighboring grains. Finally, extrusions grow rapidly and cover a high percentage of the sample surface, this corresponds to the last part after 10^8 cycles.

Another interesting point for the early damage evolution would be the implementation of the phenomenological lifetime model of Burger [65], based on the exponential growth model of Mitscherlich [66]:

$$\varepsilon_a = \left(- \frac{N_f}{K \ln\left(1 - \left(\frac{F}{F_{max}}\right)^{\frac{1}{n}}\right)} \right)^{\frac{1}{q}} \qquad \textit{Eq. 6.1}$$

Where ε_a is the strain amplitude, N_f the number of cycles to failure, F the fraction of damage, F_{max} the maximal fraction of damage, and n, K and q are fit parameters following the Basquin law from the fraction of damage as a function of the number of cycles diagram (**Fig. 6.2**b) where $N_{0.5} = K\varepsilon_a^q$.

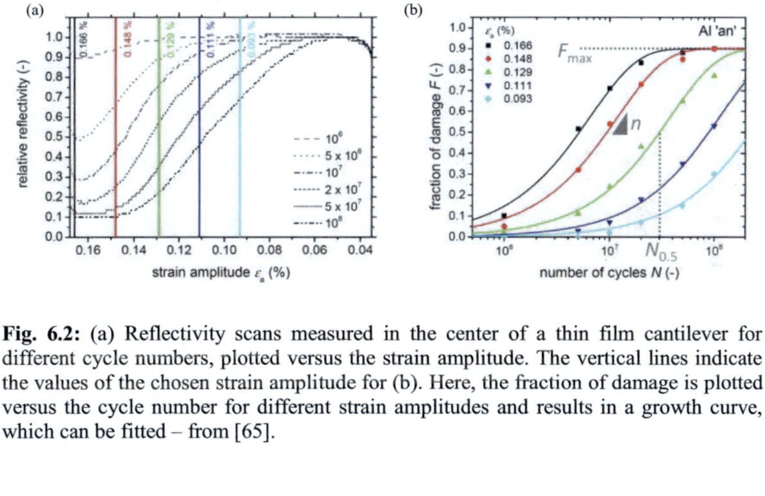

Fig. 6.2: (a) Reflectivity scans measured in the center of a thin film cantilever for different cycle numbers, plotted versus the strain amplitude. The vertical lines indicate the values of the chosen strain amplitude for (b). Here, the fraction of damage is plotted versus the cycle number for different strain amplitudes and results in a growth curve, which can be fitted – from [65].

This model could be used for the results of this thesis, where the relative black pixel amount of the whole sample (**Fig. 5.11**) or of a single grain for the very early damage (**Fig. 5.13**) would be the fraction of damage F, and where **Fig. 6.1** would be fitted to obtain the parameters n, K and q.

6.2.2 Possible early damage evolution mechanisms

To investigate the early damage evolution, different possible mechanisms have been considered in order to understand the increase of the resonant frequency observed in this thesis (e.g. in **Fig. 5.15**) during a VHCF test under a subcritical amplitude. This section presents possible explanations: cyclic hardening by the help of the DDD; vacancy formation as reviewed in section 2.5 and observed in form of pores (e.g. **Fig. 5.33**); and oxidation formation, as reviewed in section 2.3.4 and observed in **Fig. 5.4**e and **Fig. 5.6**.

6.2.2.1 Discrete Dislocation Dynamics

The group of Dr. Daniel Weygand at the KIT is specialized in Discrete Dislocation Dynamics (DDD) simulations. The DDD simulation results presented in this section were obtained by Dr. Jochen Senger during his post-doc in this group.

DDD allows for simulations of single grains or oligo grain aggregates containing dislocation structures, where complex loading can be applied [67]. To mimic the current experimental setup under high frequency bending, a surface grain is considered with a size of 2.5 x 2.5 x 2.5 μm^3. As the grain size is much smaller than the beam thickness, the stress state of the surface grain can be approximated by a uniaxial cyclic loading. To take into account that neighboring grains are not included in the model setup, cyclic bending is superimposed. Multiaxial stress states may occur in experiments even with a pure tensile stress. Indeed, strain gradients occur between single grains due to the unbalanced grain shape and the elastic anisotropy (e.g. for Cu).

For the single grain setup, the <111> direction is parallel to the tensile axis. This crystallographic orientation is chosen in agreement with the weakest orientation found with the EBSD analysis in section 5.2. The fatigued samples showed a significant defect accumulation at the surface along this orientation. In the simulation, the initial dislocation density is equal to 1.7 x 10^{12} m^{-2}, and the length of the Frank-Read source is 750 nm.

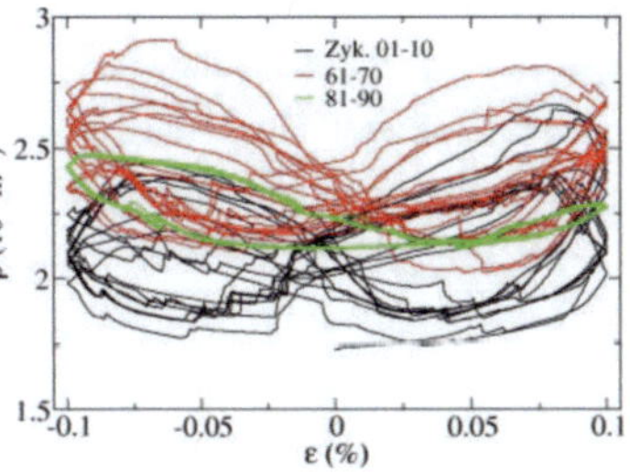

(a)

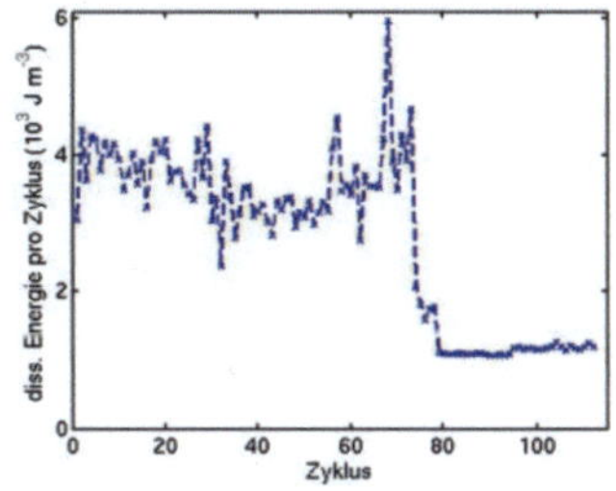

(b)

Fig. 6.3: (a) Dislocation density for different cycles in a <111> oriented grain with a volume of 2.5 x 2.5 x 2.5 µm³, (b) dissipated energy as function of cycle number.

The multiaxial stress state was found to lead to more complex dislocation networks, as cross-slip and dislocation reactions are favored. Both mechanisms introduce a higher irreversibility in the dislocation network. Furthermore, the simulations are displacement controlled and the amplitude is exceeding the usual macroscopic values used in the experiments. This can be justified by the stress and strain heterogeneity of polycrystalline samples, leading to local stresses exceeding the "mean" macroscopic values by 50% or more [68, 69]. These conditions are likely to promote local irreversible dislocation glide, which may be precursors for damage.

In **Fig. 6.3**, the dislocation density evolution for different cycles is shown. The results are displayed as sequences of ten cycles at three characteristic steps during the total number of simulated loading steps.

In the initial behavior (the first ten cycles) the dislocation density increases continuously (**Fig. 6.3**a, black curve). During each cycle the dislocation density varies, but the maximum density measured at maximal strain increases. The increase depends on the capability to form dislocation reactions leading to storage of dislocations. In **Fig. 6.3**b the dissipated energy per cycle, recorded in the DDD simulation, is plotted. It can be seen that the dissipated energy per cycle is almost constant up to cycle n°70.

Between cycles 70 and 80 the energy dissipation per cycle decreases drastically. The dislocation density and its variation shown for cycle 81-90 overlaps and the dislocation structure remains reversible (**Fig. 6.3**a, green curve). At this stage, dislocation reactions immobilized most of the dislocations. Only few dislocation segments can bow out during each cycle, but do not lead to further reactions or multiplication. The system has entered a stable state in this multiaxial loading.

With the evolution of the dislocation motion, a non-linearity of the stress state appears in this grain. At the beginning, the dissipated energy decreases significantly per cycle but in the last 30 cycles remains constant until the end of simulation (**Fig. 6.3**b). However, the analysis of the dislocation stress field in the cycle with the highest (cycle 68, **Fig. 6.3**b) and the lowest energy dissipated (cycle 90, **Fig. 6.3**b) shows only a slight difference of the respective mean values.

The analysis of the micro structural development in the 112th cycle shows the superimposed position of the dislocations. Here, the dislocation activity is limited to a single glide plane, and there is only a small number of dislocation reactions. Despite the mixed tensile / bending loading, a reversible dislocation structure has developed after 80 cycles in this grain, as other mobile dislocations have left the grain on the free surface at lower numbers of cycles.

For these small dimensions, the total number of dislocation reactions and the dislocation density are not crucial compared to the obstruction of the dislocation mobility. This dependence on single dislocations is also observed in bending simulation of monocrystalline beams [70].

To better understand the impact of a superimposed bending moment, another simulation with the previous structure (**Fig. 6.4**a) has been conducted, where the bending angle is increased from ± 3.2° to ± 4.8°. By increasing the bending angle, the strain gradient increases. This generates more geometrically necessary dislocations. A higher dislocation density leads to a higher probability of junction formation, cross-slip and therefore also dislocation multiplication. Increasing the multiaxiality increases the formation of small prismatic dislocation loops (**Fig. 6.4**b). Because of their small size, the shear stress gradients are too small to allow for gliding of prismatic loops. Again, fewer dislocations are mobile and there are more and more obstacles hindering glide. This results in a smaller variation in the dislocation density per cycle (**Fig. 6.4**c, black curve). Thus a smaller plastic deformation takes place in the grain at higher cycles (**Fig. 6.4**d). The cyclic hardening leads to a decrease of the dissipated energy (**Fig. 6.4**e).

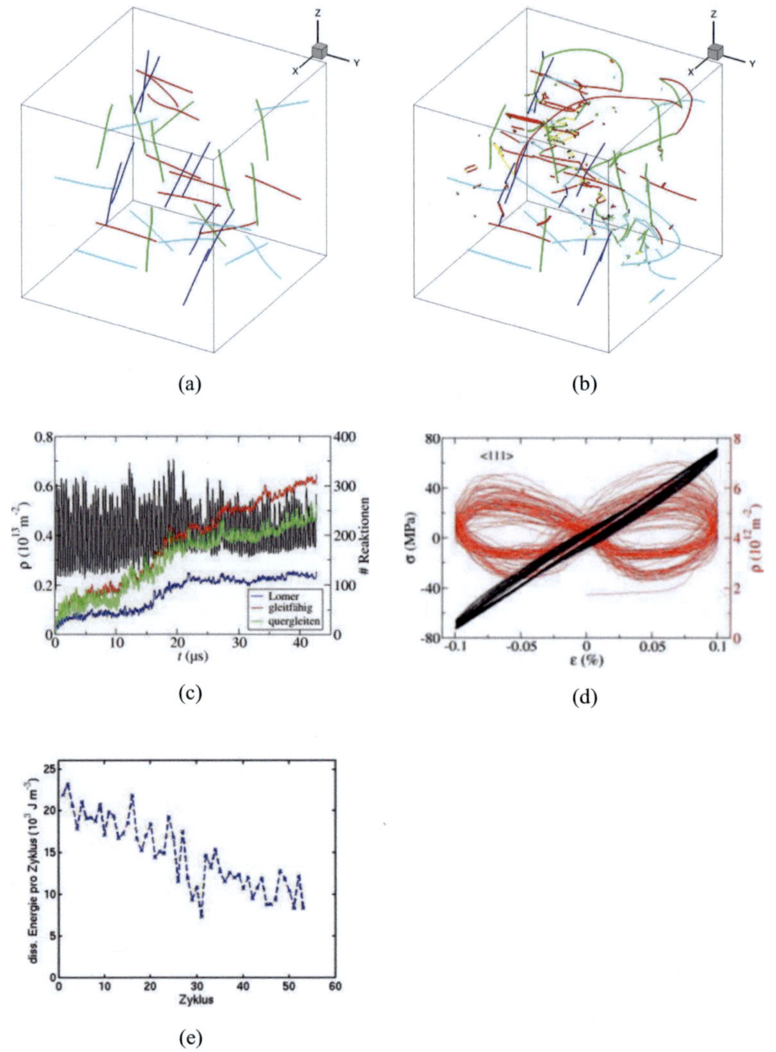

Fig. 6.4: <111> -oriented grain: (a) initial structure consisting of Frank-Read sources, (b) dislocation structure after 52 cycles. (c) Dislocation density and number of reactions as function of time. (d) stress-strain curve with the corresponding dislocation density development. (e) Dissipated energy as function of cycles. In this simulation, only the YZ surface at x_{max} is permeable for dislocations.

With increasing irreversibility of the dislocation structure, favored by increased multiaxiality (increased bending moment) a rougher surface is observed. However, the observed irregularities due to emerging dislocations in the achieved number of cycles are still very small. The change in height is less than the absolute value of a Burgers vector.

6.2.2.2 Frequency increase – subcritical amplitude – cyclic hardening behavior

The resonant frequency is controlled by the elastic properties of the material as well as the geometry of the sample. It is assumed that the geometry of the sample is not changing during fatigue for a subcritical amplitude. To get a better insight, hysteresis curves of the DDD results were analyzed with respect to an anelastic behavior. These results have been obtained by the group of Dr. Daniel Weygand using DDD simulations (see section 6.2.2.1).The anelastic behavior stems from a bowing out of the dislocation. The cyclic dislocation motion is associated with phonon dissipation and therefore leads to a hysteresis. In a rather simple approximation, the stresses values at maximal strain were assumed to be of pure elastic origin, therefore defining an effective Young's modulus. This modulus is normalized by the modulus of the first cycle (relative Young's modulus). **Fig. 6.5**a shows the relative Young's modulus as function of the number of cycles. It was observed that after initial softening (dislocation multiplication) and after 90 cycles the relative module is increased and reaches a roughly 1% higher stiffness than at the beginning of the simulation. This cyclic hardening has been explained in the previous section (**Fig. 6.4**d) by dislocation immobilization. Here dislocation reactions lead to the pinning of the mobile dislocations and therefore the energy dissipation is reduced. This smaller dissipation leads to a smaller stress-strain hysteresis, leading to a higher stiffness.

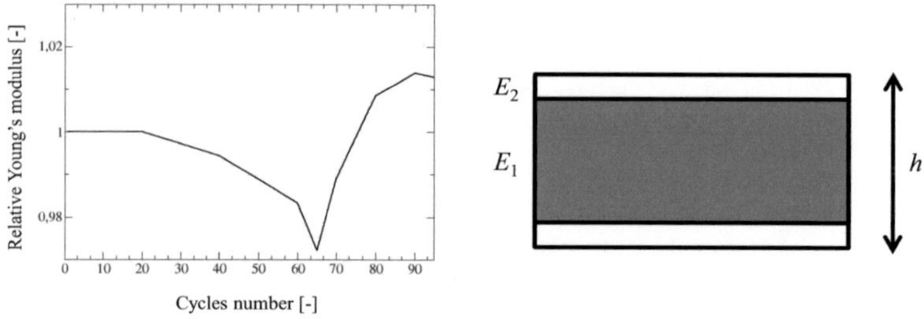

(a) (b)

Fig. 6.5: (a) Cyclic hardening after 65 cycles: increase of the relative module, (b) anelastic behavior mechanism with a larger Young's modulus for the grains at the surface.

It is assumed that only the Young's modulus of the outer layer E_2 varies. Eq. 6.2 expresses the resonant frequency f as function of the Young's modulus E of the sample.

$$f = \frac{1}{2\pi}\sqrt{\frac{k}{m}} = \frac{1}{2\pi}\sqrt{\frac{3EI_y}{ml^3}} = \frac{1}{2\pi}\sqrt{\frac{3I_y}{ml^3}}\sqrt{E} \qquad \text{Eq. 6.2}$$

In order to calculate the frequency change as function of the Young's modulus change, Eq. 6.2 has been differentiated:

$$\Delta f = \frac{1}{2\pi}\sqrt{\frac{3I_y}{ml^3}}\frac{\Delta E}{2\sqrt{E}} \qquad \text{Eq. 6.3}$$

Finally the relative frequency change can be obtained:

$$\frac{\Delta f}{f} = \frac{1}{2}\frac{\Delta E}{E} \qquad \text{Eq. 6.4}$$

The following geometrical assumptions have been made:

- The core material is unchanged and has a Young's modulus of E_1,
- 10% of the total sample thickness (corresponding to a layer of grains at the top and bottom surface) has a 1% higher Young's Modulus E_2.

The global Young's modulus of the mechanism is roughly equal to:

$$E = 0.9\,E_1 + 0.1\,E_2 \qquad \textit{Eq. 6.5}$$

The DDD simulations suggested (**Fig. 6.5**a) that E_2 increases by 1%:

$$\frac{\Delta E_2}{E} = 10^{-2} \qquad \textit{Eq. 6.6}$$

Since E_1 is constant, the differential of Eq. 6.5 gives:

$$\frac{\Delta E}{E} = 0.1 \text{ x } \frac{\Delta E_2}{E} = 0.1 \text{ x } 10^{-2} = 10^{-3} \qquad \textit{Eq. 6.7}$$

Finally, the relative frequency change corresponding to the anelastic behavior of this mechanism is equal to:

$$\frac{\Delta f}{f} = 5.10^{-4} \qquad \textit{Eq. 6.8}$$

This value of the relative frequency increase after only 90 cycles has the same order of magnitude as what can be observed in **Fig. 5.15** for the really first cycles. Therefore, cyclic hardening is a possible reason for a resonant frequency increase up to the VHCF regime.

6.2.2.3 Frequency increase – subcritical amplitude – vacancy formation

The second explanation for the resonant frequency increase takes into account an increase of the sample thickness due to a vacancy formation in the outer layer of the sample (**Fig. 6.6**). This possible mechanism has been reviewed in section 2.5 and observed in form of pores (e.g. **Fig. 5.33**)

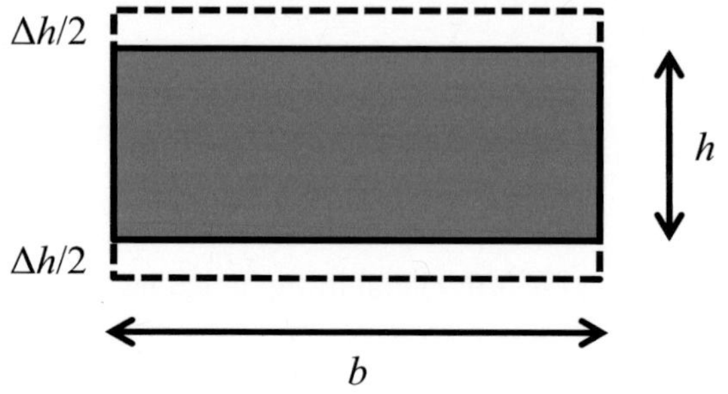

Fig. 6.6: Vacancy formation mechanism involving an increase of the sample thickness.

The resonant frequency f has been expressed as function of the sample thickness h:

$$f = \frac{1}{2\pi}\sqrt{\frac{k}{m}} = \frac{1}{2\pi}\sqrt{\frac{3EI_y}{ml^3}} = \frac{1}{2\pi}\sqrt{\frac{3b}{12ml^3}}\sqrt{h^3} \qquad \textit{Eq. 6.9}$$

The same method as for the previous mechanism is used. The resonant frequency differential is equal to:

$$\Delta f = \frac{1}{2\pi}\sqrt{\frac{3b}{12ml^3}}\frac{3}{2}\frac{h^2}{\sqrt{h^3}}\Delta h = \frac{1}{2\pi}\sqrt{\frac{3b}{12ml^3}}\sqrt{h^3}\frac{3}{2}\frac{\Delta h}{h} \qquad \textit{Eq. 6.10}$$

It can be concluded that the relative frequency change is equal to:

$$\frac{\Delta f}{f} = \frac{3}{2}\frac{\Delta h}{h} \qquad \textit{Eq. 6.11}$$

The following simplifications are used:

- In a single grain, a vacancy (of the size of the Burgers vector (e.g. |b| = 0.2862 nm for Al) is created per cycle, which corresponds after 109 cycles to a volume change of ΔV = 2.3 x 107 nm3,
- the change of the relative grain volume (for a cubic grain size of 15 µm) is equal to the change of the relative sample thickness (for a thickness of 200 µm), as shown in **Fig. 6.6**:

$$\frac{V + \Delta V}{V} = \frac{h + \Delta h}{h} \qquad \textit{Eq. 6.12}$$

The volume of a cubic grain with a length of 15 μm is equal to:

$$V = 3.38 \text{ x } 10^{-15}\, m^3 \qquad \textit{Eq. 6.13}$$

Δh can be calculated from Eq. 6.12:

$$\Delta h = 1.39 \text{ x } 10^{-9}\, m \qquad \textit{Eq. 6.14}$$

Finally, the relative frequency change corresponding to a vacancy formation is:

$$\frac{\Delta f}{f} = 1.04 \text{ x } 10^{-5} \qquad \textit{Eq. 6.15}$$

This relative resonant frequency increase appears when a vacancy is formed in a single grain per cycle after 10^9 cycles. This value is rather low compared to the results shown in section 5.1.

One could argue that vacancies appear in more than just a single grain (e.g. ten grains), which would increase the total amount of the relative resonant frequency by a factor of 10. This would still be really low compared to the relative frequency increase of about 2 x 10^{-3} observed in the experiments.

One could also think about looking at the volume change due to subsurface pores (probably intensively higher than what has been taken into account here), such as in **Fig. 5.24**b, and calculate the corresponding resonant frequency increase.

6.2.2.4 Frequency increase – subcritical amplitude – oxidation formation

As explained in section 2.3.4, one can expect that cracks may grow because of fatigue-induced surface oxide mechanism.

Boyce et al. [18] observed a 400 nm in thick NiO layer (**Fig. 2.9**d), which is several orders of magnitude thicker than the native oxide. TEM observations of Baumert et al. [19, 20] revealed localized surface oxides as thick as 700 nm (**Fig. 2.10**).

This last mechanism investigates the influence of an oxidation layer on the relative resonant frequency.

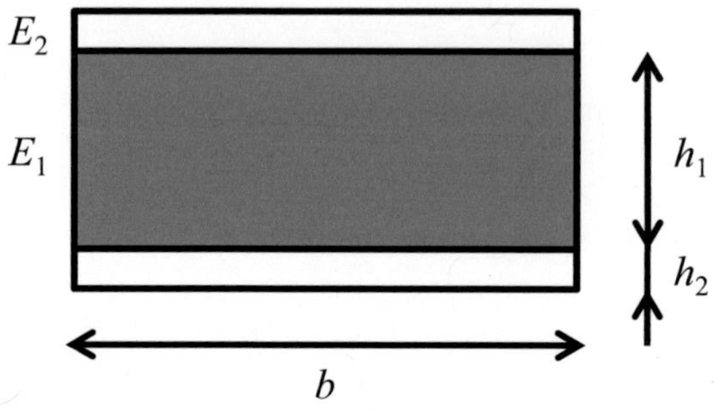

Fig. 6.7: Oxidation mechanism with an oxidation layer increasing the sample thickness.

Since an oxide layer has an influence on the sample stiffness and on the whole Young's modulus, the sample resonant frequency f has first been expressed as function of the sample stiffness k:

$$f = \frac{1}{2\pi}\sqrt{\frac{k}{m}} = \frac{1}{2\pi}\sqrt{\frac{1}{m}}\sqrt{k} \qquad \textit{Eq. 6.16}$$

The differential of Eq. 6.16 is:

$$\Delta f = \frac{1}{2\pi}\sqrt{\frac{1}{m}}\frac{1}{2}\frac{\Delta k}{\sqrt{k}} \qquad \textit{Eq. 6.17}$$

And the relative frequency change as function of the sample stiffness is equal to:

$$\frac{\Delta f}{f} = \frac{1}{2}\frac{\Delta k}{k} \qquad \textit{Eq. 6.18}$$

In this case, the stiffness k is determined by the thickness of the sample h_1 and its Young's modulus E_1 and the oxidation layer thickness h_2 and Young's modulus E_2.

The following assumptions have been made:

- An oxidation layer of 400 nm [18] grows only on the damaged surface grains in the middle of the sample and therefore do not change the mass at the end of the sample.
- The oxide layer only grows on the top and bottom surface of the sample where the stress amplitudes are maximal. It is assumed that the sample sides are free of oxide.
- It is assumed that on average 10% of the width is damaged and has an oxidation layer of 400 nm, see **Fig. 5.21** and **Fig. 5.25**, the equivalent thickness increase of the whole width is equal to 40 nm.
- An aluminum sample with a width of b = 500 µm, with h1 = 250 µm and E1 = 70 GPa, and a layer of aluminum oxide Al2O3 with h2 = 40 nm and E2 = 350 GPa has been chosen as material geometries and parameters. Aluminum maximizes the impact of oxide since Al2O3 has an enormous impact on the global Young's modulus. In case of a Ni-sample the effect of an oxide would be much less.

The initial stiffness is equal to:

$$k_{ini} = E.I = E_1.I_1 = E_1.\frac{b.h_1^3}{12} \qquad \textit{Eq. 6.19}$$

Analytically $k_{ini} = 4.56 \times 10^{-5}$ N/m.

The effective stiffness, including the oxidation layer, has to be calculated as function of h_1, h_2, E_1 and E_2:

$$k_{eff} = E.I = E_1.I_1 + E_2.I_2 \qquad \textit{Eq. 6.20}$$

$$k_{eff} = E_1.\frac{b.h_1^3}{12} + E_2.b.\left[\frac{h_2^3}{6} + \frac{h_2.(h_1 + h_2)^2}{2}\right] \qquad \textit{Eq. 6.21}$$

Knowing that:

$$\Delta k = k_{eff} - k_{ini} \qquad \textit{Eq. 6.22}$$

It is possible from Eq. 6.18 to calculate the relative frequency change due to an oxidation formation:

$$\frac{\Delta f}{f} = 2.4 \text{ x } 10^{-3} \qquad \textit{Eq. 6.23}$$

This relative resonant frequency increase is consistent with the one observed in the experimental section 5.1.2.

Due to oxide formation observation in this work (**Fig. 5.4**e) and possibly in **Fig. 5.6**, and due to the same order of magnitude between the calculated and the observed increase of the resonant frequency, this mechanism seems to be a plausible explanation for the resonant frequency increase up to the VHCF regime.

It has been shown before that fatigue-induced surface oxides act as a catalyst for micro-crack nucleation (section 2.3.4) and that as soon as micro-cracks are initiated the resonant frequency drops. One could think of a first mechanism where a thin oxide layer grows on damaged grains and increases the resonant frequency. But as soon as slip bands are large enough to activate the typical extrusion/intrusion mechanism, micro-cracks nucleates and the resonant frequency decreases. This may correspond to the results of the Cu34 test in section 5.3.3.3, where a crack initiation mechanism has probably been activated in the VHCF regime. Other reasons for this late crack initiation are discussed in the section 6.3.1.

6.2.3 Outlook: high-throughput method

The implementation of a novel high-throughput method for the investigation of damage evolution from before the crack initiation to the short crack growth is work in progress. This method would be based on the very successful new custom-built resonant fatigue setup. The idea is to add an in-situ optical acquisition system and an on-line image processing methodology. One would observe individual grains and their neighbors on the whole sample surface.

To get statistically significant data, a high-throughput method would be preferable. The micro-bending fatigue method described in this work induces a distribution of stress amplitudes along the sample. The local stress state can be correlated to the position of the individual grains. Thanks to an automated image evaluation, based on many years of work on image analysis in the group of Prof. Eberl, damage evolution can be recorded, tracked, and evaluated for these damaged grains and their neighbors. Through EBSD, the grain orientation of all grains can be known before test. And at the same time, the relationship between the optically observable changes on the sample surface and the mechanical stiffness change can be detected by quantitatively looking at the relative frequency change of the sample.

This method gives in-situ insights of the damage evolution distribution in individual grains and neighbors, depending on the number of cycles, the local stress amplitude, and the grain orientation.

6.3 Crack initiation in fcc materials

6.3.1 Crack initiation mechanism in the VHCF regime

With the custom-built resonant fatigue setup developed in this work, it is possible to observe at the beginning of an experiment, if either a decrease or an increase of the resonant frequency will occur.

If a subcritical amplitude is applied, an increase of the resonant frequency is observed (**Fig. 5.15**). The investigation on damage initiation conducted by SEM and laser scanning microscope showed that in many cases typical damage mechanism (roughness, slip bands, pores or micro-cracks) occurred but only in a few grains.

However, if an overcritical amplitude is chosen, a decrease of the resonant frequency of the sample, associated with a unique microstructural damage development, can occur. It can also be observed that after more than 10^7 cycles, the slope of the decrease is reduced; the damage evolution is slower and can even saturate.

This behavior could be used as a fast fatigue test by observing the resonant frequency response while the stress amplitude is increased step by step. At a chosen amplitude, if the resonant frequency decreases, the sample will definitely fail. On the contrary, if the resonant frequency increases under constant stress amplitude, the sample may not fail within 10^9 cycles. Nevertheless it is not clear how fast the damage mechanisms can or have to develop as a function of the stress amplitude. And even if the resonant frequency is increasing at the beginning of the test, it may decrease later on. In this work for example, the damage in the VHCF for Al and Cu looks very similar and is comparable with what Höppel et al. [71] observed. For all tests where the relative resonant frequency increased with small stress amplitudes, SEM observations showed slip bands on only a few isolated grains with plastically undamaged neighbors.

One can fairly assume that this was the case for the first part of the Cu34 test (section 5.3.3.3), where the resonant frequency increased. Nevertheless, the resonant frequency of Cu34 decreased after $N_f = 3.99 \times 10^8$ cycles. It can be considered that another mechanism has been initiated after such a long time. Some arguments can help the understanding of this mechanism and open the topic for discussion.

As used in the FEM model of Krupp et al. [29], one could think that the stored energy by the accumulated dislocation pile up (accumulation of irreversibilities [8], see section 2.5) reached the critical value of the surface energy, and that the dislocation dipoles transform in free surface, corresponding to the crack initiation at slip bands, as reviewed in section 2.3.3 and section 2.4.

Another explanation could be that even under low stress amplitudes a high number of cycles allows for the atmospheric oxygen enough time to react with the underlying base metal, and thereby activate an oxide thickening mechanism (Boyce et al. [18], see section 2.3.4), through which the fatigue-induced surface oxide reached a thickness large enough to initiate cracks at slip bands.

However, if this energy threshold is not reached, even if some non-negligible damage has already appeared, micro-cracks will not be able to initiate or propagate, and the sample will not fail. Nowadays, there is no proof that the initiation of such a mechanism could not appear in the late VCHF regime (e.g. above 10^{10} cycles). And therefore, it is challenging to give an interpretation about a VHCF endurance limit for fcc materials.

6.3.2 Experimental input for crack initiation model

6.3.2.1 Tanaka-Mura, and Chan's crack initiation model

As reviewed at the end of section 2.4.4, and described in section 6.3.1, a possible way to model the early damage formation and specially the crack initiation is to consider an energy balance.

Once the stored energy by the accumulated dislocation pile up equals the surface energy, the dislocation dipoles transform into a free surface. This corresponds to the crack initiation. The necessary condition for the initiation of a crack with a length a at a number of cycles N_i can be written according to [29] as:

$$N_i = \frac{4\mu^2 h^2 a}{\lambda\pi(1-\nu)(\Delta\tau - 2k)^2 d^3} \qquad \textit{Eq. 6.24}$$

Where:

- μ is the shear modulus,
- h the slip band width also called slip band spacing,
- a the crack length,
- λ the cyclic irreversibility ($0 < \lambda < 1$),
- ν the Poisson ratio,
- d the grain size,

- $\Delta\tau$ the shear-stress range, with $\Delta\tau = \Delta\sigma \cos\varphi \sin\theta$, where $\Delta\sigma$ is the range of the local applied stress, φ is the angle between the normal of the slip plane and the direction of the applied force, and θ is the angle between the slip direction and the direction of the applied force.
- k the critical shear strength considered as cyclic frictional shear stress that is required to move dislocations along a slip plane.

6.3.2.2 Experimental measurement

The new custom-built resonant fatigue setup presented in this work can be used to measure the parameters needed for the Chan's crack initiation model [21].

In this model, some material parameters such as the shear modulus μ and the Poisson ratio v are known. Some other parameters such as the shear-stress range $\Delta\tau$ and the critical shear strength k are depending on experimental conditions. The sample microstructure with the gain size d also plays a role, and can be found e.g. by using EBSD.

Finally some of the parameters have to be characterized experimentally. This is where the new custom-built resonant fatigue setup can be used to find the relevant parameters needed for different models:

- Slip bands can be detected and observed as seen in **Fig. 5.2**. At this step, the slip band width h can easily be measured with an AFM or a laser scanning microscope.
- The evolution of the crack length a can be observed by the resonant frequency change, as small as hundreds of nm (e.g. **Fig. 5.3**).
- Finally, the cyclic slip irreversibility λ can be found in the literature for few materials and experimental conditions, and only under certain assumptions. Mughrabi gathered some results (**Tab. 6.1**) on the cyclic slip irreversibilities in his review [8].

Example	Testing conditions	Irreversibility λ
PSBs in copper	$\gamma_{pl,PSB}\approx10^{-2}$, RT	≈0.3
VHCF, copper	$\gamma_{pl,PSB}\approx1.7\text{x}10^{-5}$, RT	≈0.000034
α-brass polycrystals	$\Delta\sigma/2$=82 MPa, RT	≈0.005
	$\Delta\sigma/2$=82 MPa, RT	≈0.03
Cu-2 at. % Co polycrys-	$\gamma_{pl,PSB}\approx2.3\text{x}10^{-2}$, RT	≈1
tals	$\Delta\varepsilon_{pl}/2=10^{-3}$, 650°C	≈0.05
Waspaloy	$\Delta\varepsilon_{pl}/2=5\text{x}10^{-4}$, RT	≈0.03
α-iron polycrystals	$\Delta\varepsilon_{pl}/2=5\text{x}10^{-4}$, RT	≈0.06
	$\Delta\varepsilon_{pl}/2=5\text{x}10^{-4}$, RT	≈0.2

Tab. 6.1: List of examples of different fatigued materials, for which the cyclic slip irreversibility could be estimated – from [8].

This parameter can nevertheless be qualitatively found with the new custom-built resonant fatigue setup, looking at the evolution of the resonant peak during the test. Indeed if no cracks are initiated on the sample surface, a wider resonant peak, corresponding to a higher damping, implies cyclic irreversibilities. The damping can be evaluated with the quality factor, corresponding to the peak (**Fig. 6.8**a) width at the half maximum amplitude. It can also be quantified with the slope of the curve of the phase angle (**Fig. 6.8**b) as function of the frequency, decreasing with damping.

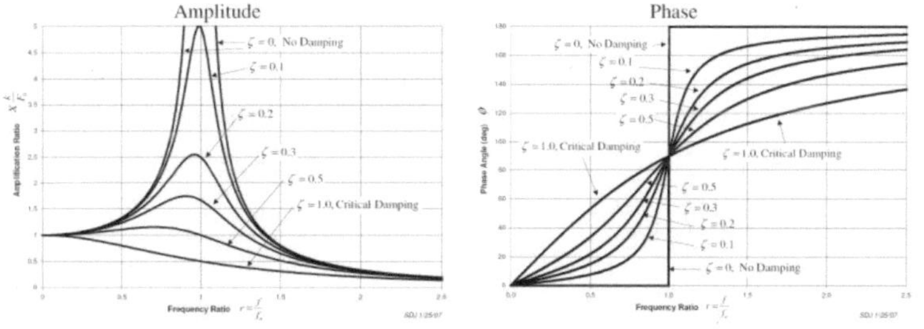

(a) (b)

Fig. 6.8: Relative response magnitude (a) and phase (b) with damping factors (ζ) of 0 (no damping) to 1.0 (critically damped) – from https://en.wikipedia.org/wiki/Vibration.

Moreover, the cyclic slip irreversibility parameter λ can also be quantitatively evaluated.

According to Mughrabi [8], the evolution of surface roughness in the PSB after *N* cycles can be described by the mean width *w*; defined as the largest peak-to-valley spacing, measured parallel to the Burgers vector, as:

$$w = \sqrt{6bh\lambda N \Delta\gamma_{pl}/2} \qquad \textit{Eq. 6.25}$$

where *b* the length of the Burgers vector **b,** *h* denotes the thickness of the glide region studied, λ the cyclic slip irreversibility, and $\Delta\gamma_{pl}/2$ the shear strain amplitude.

The cyclic slip irreversibility can be obtained from Eq. 6.25:

$$\lambda = \frac{w^2}{6bhN\Delta\gamma_{pl}/2} \qquad \textit{Eq. 6.26}$$

Where *w*, and *h* can easily be measured on the sample surface after *N* cycles. The cyclic plastic shear strain amplitude $\Delta\gamma_{pl}/2$ can be calculated from the plastic strain amplitude $\Delta\varepsilon_{pl}/2$ which can be estimated with different testing methods: e.g. by thermoelectric methods, infrared camera and strain gage mcasurements of hysteresis area, as explained for 19 kHz ultrasound tests in the Stanzl-Tschegg review [72].

These methods are nevertheless not appropriate for the micro-samples presented in this work. The sample temperature increase is too low due to a high surface to volume ratio, and therefore is extremely difficult to measure. It is finally highly complicated to fix a strain gauge on the sample due to its size and its displacement in test.

One possibility for calculating the cyclic plastic strain $\Delta\varepsilon_{pl}/2$ is to look at the slope of the phase angle $\Delta\varphi/\Delta f$ (**Fig. 6.8**b) around 90° as function of the total strain. As soon as irreversibilities appear in the sample, the slope decreases due to damping. The same observations can be made when looking at the sample amplitude over the piezo amplitude ratio, representing the dissipated energy between the sample and the piezo actuator (**Fig. 6.9**).

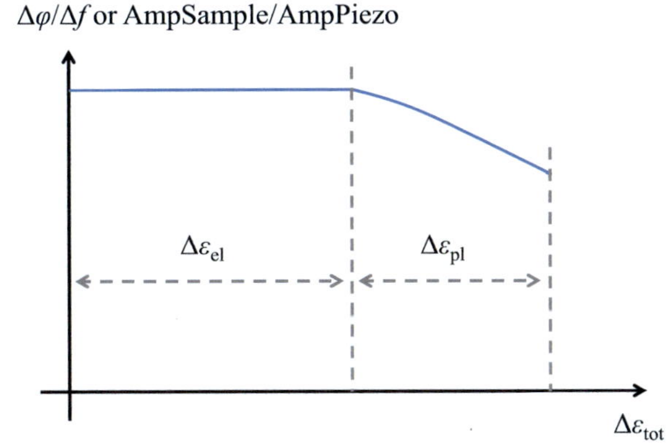

Fig. 6.9: Slope of the phase angle $\Delta\varphi/\Delta f$ and sample amplitude over the piezo amplitude ratio as a function of the total strain.

As a conclusion, the new custom-built resonant fatigue setup coupled with fatigue damage observations can be used for measurements of necessary parameters for crack initiation models. Chan's model could be implemented in the future work on the VHCF behavior of small scale samples.

6.3.3 Comparison of aluminum fatigue lifetime

The following section shows a comparison of the fatigue results of Al with literature values from Thompson and Backofen [73] and Höppel et al. [71]. Finally, the influence of the material purity on the fatigue behavior is investigated.

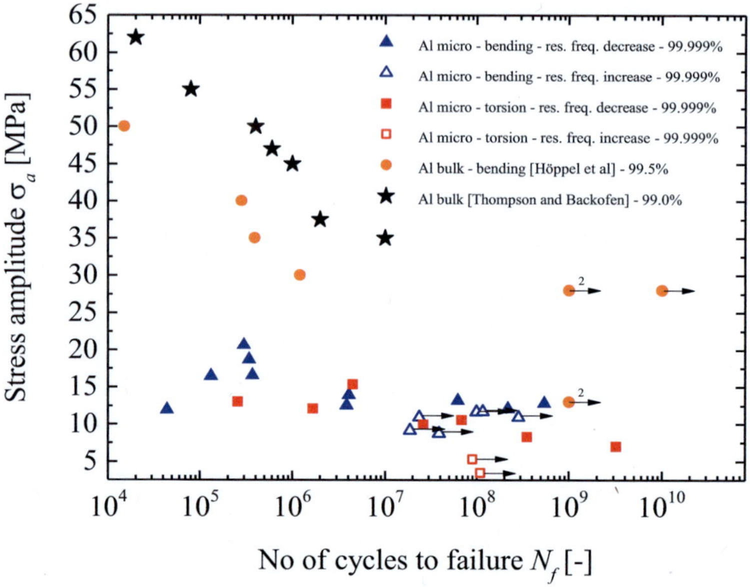

Fig. 6.10: Pure Al lifetime diagram.

Fig. 6.10 shows the fatigue lifetime of pure Al micro-bending samples compared to bulk Al samples. The micro-bending and torsion experiments from this work use a failure criterion corresponding to a change of the relative resonant frequency ($\Delta f / f_{\text{initial}} = -2 \times 10^{-3}$) when cracks start to initiate. Indeed for micro-samples the VHCF regime is governed by crack initiation. This is why the different lifetimes presented in **Fig. 6.10** can be compared.

However the fatigue lifetime of these 500 x 250 µm² aluminum samples show a smaller lifetime than the one from Thompson and Backofen or Höppel et al.. This can be explained through the different material purities. **Tab. 6.2** lists the different impurities in percentage of the mass for the different aluminum samples from this comparison.

	Principal impurities in % of the mass					
Material	**Fe**	**Pb**	**Zn**	**Si**	**Mg**	**Cu**
AA 1100 99.0% (Thompson et al.)	0.083	<0.01	0.003	0.120	-	0.140
AA 1050 99.5% (Höppel et al.)	0.239	0.044	0.060	0.175	0.040	
Al 000630 99.999% (This work)	0.00003	-	-	0.00008	0.00012	0.00003

Tab. 6.2: Purity of the different Al compared in this study.

The 1100 aluminum alloy specimens of 2.5 cm gauge length and 0.8 cm diameter with a grain size between 20 µm and 200 µm were fatigued (fully reversed loading at 30 Hz) by Thompson and Backofen [73]. The samples were machined to the final dimensions, and then mechanically and chemically polished before heat treatment. Thompson and Backofen [73] conducted different tests until the end of the HCF regime (10^7 cycles). The Al S-N curves for different grain sizes converge after 10^5 cycles.

They explained that for Al (and also Cu) samples, different grain sizes between 20 and 200 µm do not have an influence on the lifetime due to the fact that cross-slip is easy in such material. Indeed they argued that the micro structure in fatigued aluminum sample is masking the grain size.

Höppel et al. [71] investigated aluminum of 99.5% purity (AA1050). The microstructure has a coarser structure after recrystallization state and has a globular grain geometry with an average grain size of 55 µm. Rods of 12 mm in diameter and 100 mm length were cut out of the material and each sample was mechanically and electrolytically polished in order to obtain a high quality surface. Höppel et al. used a Roell Amsler vibrophore system HFP 5100 for HCF tests at about 130 Hz. The cyclic deformation tests were performed under stress control using a stress ratio of R = -1. In cooperation with the laboratory of Prof. Christ, University of Siegen, Germany, additional VHCF tests were performed at stress amplitudes of 13 MPa and 28 MPa, in the range of 10^9 up to 10^{10} cycles. The ultrasonic testing frequency was conducted at 20.65 kHz. The usual ultrasonic procedure has been used for the tests: to avoid undesired heating of the specimens the fatigue tests were performed using air-pressure cooling and pulse-pause-sequences, where each pulse sequence lasts 300 ms.

Höppel et al. [71] showed that 99.5% Al can exhibit a classical fatigue limit. The authors point out that for all specimens which failed, the cracks started at the surface. For the stress amplitude of 28 MPa (which is slightly below the fatigue limit of their material) and of 13 MPa, no failure was detected in the VHCF regime. Nevertheless, they observed that for these stress amplitudes intense surface roughening and pronounced cyclic slip markings are visible. They also remarked that with decreasing stress amplitude the density of cyclic slip markings on the surface becomes smaller and the plastic deformation becomes localized in particular grains, while neighboring grains remain plastically undeformed. They stated that surface roughening, as a measure of slip irreversibility, takes place in the VHCF regime although the stress amplitudes are (significantly) below the classical fatigue limit.

It is of high importance at this point to specify that the case of Cu34 (see 5.3.3.3) brings more insights to the VHCF mechanism observed by Höppel et al. Further explanations are given in 6.2.2.4.

Finally, the influence of purity on the fatigue lifetime has been investigated. Thompson and Backofen worked with a 99.0% aluminum alloy where the samples have the highest fatigue lifetime, Höppel et al. a 99.5% aluminum alloy with a lower lifetime, while this work presents the results of a 99.999% pure aluminum, from which the samples have the lowest lifetime.

The review on aluminum alloys from Mondolfo [74] shows the same tendency. As can be seen in **Fig. 6.11**, at 298 K, a 99.99% pure aluminum has a lower lifetime than one with a 99.98% purity. The same observation can be made at 293 K for a 99.99% pure aluminum and a 99.2% pure one.

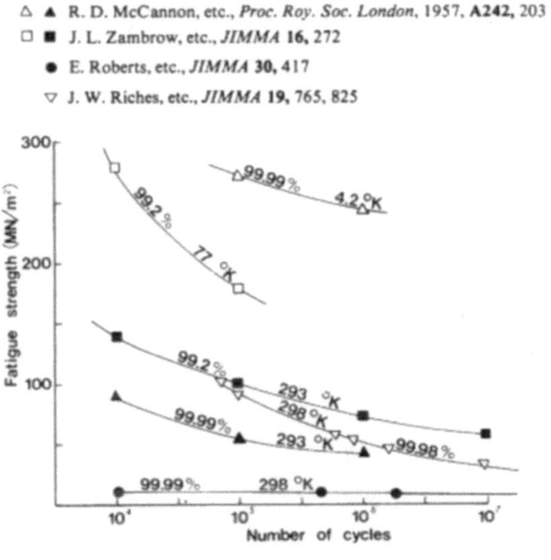

Fig. 6.11: Fatigue of aluminum as function of purity and temperature, according to various authors – from [74].

Mondolfo concluded that most factors that increase the ultimate tensile strength also increase the fatigue resistance: cold work, low temperature, and alloying, which matters in this study. All of these factors actually improve the endurance limit.

The author finally stated that in commercial metals, the fatigue strength increases with iron and silicon content. This can be observed for both the AA1050 and the AA1100 in **Tab. 6.2**, and explains why they have a higher fatigue lifetime compared to the 99.999% Al micro-samples from this work (**Fig. 6.10**).

6.3.4 Comparison of OFHC copper fatigue lifetime

In this section, the results from the experimental investigation on Oxygen-Free High Thermal Conductivity (OFHC) Cu micro-bending samples are compared to bulk material samples and thin films.

It shows that dimensional constraints cause size effects in small-scale samples (**Fig. 6.12**). It has to be noted though, that the comparison between the experimental lifetime results should be reviewed critically due to the different lifetime criteria.

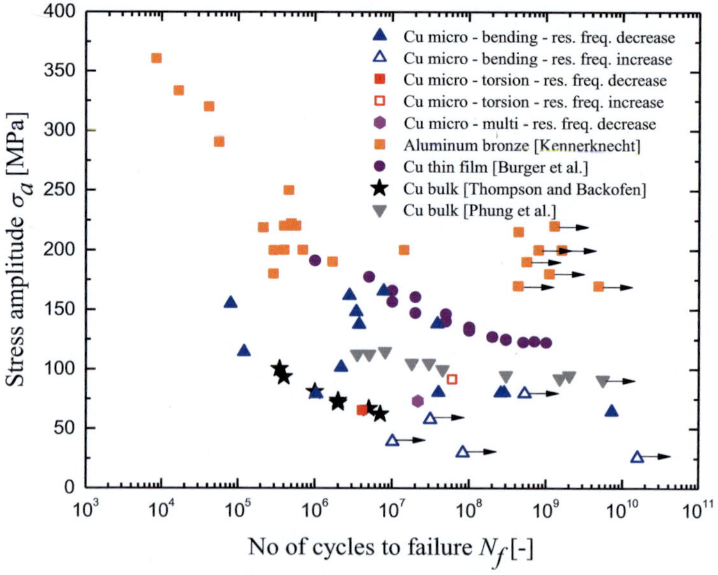

Fig. 6.12: OFHC Cu lifetime diagram.

On the macro scale, OFHC copper cylindrical specimens of 2.5 cm gauge length and 0.8 cm diameter with a grain size between 3.4 µm and 150 µm were reported (fully reversed loading at 30 Hz) by Thompson and Backofen [73]. They explained that for Cu (and as already explained for Al) samples, the different grain sizes do not have an influence on the lifetime due to the fact that cross-slip is easy in such material. OFHC Cu (99.95%) with a grain size of 30 µm was also studied by Phung et al. [40] with fully reversed loading at 20 kHz. Phung et al. used cylindrical and flat samples having gauge lengths of about 3 cm, with the cylindrical samples having diameters of 3 mm, and the flat samples a cross section of 3 mm x 2 mm.

The difference between the lifetime measured by Thompson and Backofen and Phung et al. could be due to the preparation of the OFHC Cu, or other experimental differences, a notable difference being the extremely different test frequencies (30 Hz for Thompson and Backofen, 20 kHz for Phung et al.). In both cases the samples were fatigued until failure. In contrast, for the micro-bending fatigue tests from this study a failure criterion related to the decrease of the relative resonant frequency ($\Delta f / f_{\text{initial}} = -2 \times 10^{-3}$) was used. As described in the sections before, this frequency change is associated with crack initiation and growth. It can be argued that in the higher HCF and the VHCF regime, the practical lifetime of micro-samples is mainly determined by crack initiation. Therefore, the two lifetime criteria will lead to similar fatigue lifetimes in that regime. The fatigue lifetime of 500 x 200 µm^2 OFHC Cu specimens measured with the custom-built resonant setup show altogether a higher lifetime compared to values from Thompson and Backofen for a comparable grain size. Although the lifetime data shows significant scatter, the lower lifetime limit of the scatter is the same as for macroscopic samples. Small samples, such as most of the micro OFHC Cu specimens, are therefore not large enough to contain a representative microstructure of the bulk material, and show a widely distributed lifetime. Consequently, micro-samples have on average a higher lifetime, but carry a high risk to fail.

Comparing these results, it can be concluded that the lifetime of the Cu micro-samples is controlled by a statistical size effect. For the bending fatigue tests of Cu micro-samples, the highest stressed volume is limited to only a couple hundred grains. Hence, it can be argued that these samples are not representative elements of the natural defect distribution. Nevertheless, such non-representative distributions are often used in applications such as micro components cut out of a macro material. Such samples are also useful for testing a targeted microstructure and investigating specific fatigue mechanisms. Finally, in terms of the Cu micro-samples, the probability is low that arrangements of grains are present which can lead to damage accumulation crossing several grains, resulting in fatal cracks [41].

This is similar to the findings of Furuya [45] who has shown that the sample size affects the lifetime of high strength steel in the VHCF regime. Furuya's results show that the scatter of the sample lifetime increases with decreasing sample size, but only after the samples become smaller than a certain critical volume.

The initial defect distribution is typically related to the manufacturing or deposition process, and the subsequent thermal treatment. The sample lifetime can be influenced by changing the manufacturing parameters. This is well illustrated by the uniaxial fatigue tests with micro molded tensile samples (cross section 260 x 130 μm², gage length 1 mm) made of aluminum bronze (CuAl10Ni5Fe4) [75]. The lifetime of the CuAl10Ni5Fe4 micro-samples in contrast shows no scatter at all. This seems inconsistent with Furuya's argument, but the reason for this behavior can be found in the very fine microstructure resulting from the manufacturing process. While the Cu micro-samples were extracted from a large Cu foil, the CuAl10Ni5Fe4 samples were micro-molded directly into the final shape [76]. Therefore, the samples contain all defects and microstructural features developed during the molding process and are representative elements for the characteristic defect distribution.

These results can also be compared with thin film fatigue tests to further explore the size effects. 1 μm Cu thin films were sputter deposited on the Si cantilevers with dimensions of 10 x 1 x 0.2 mm^3, and the cantilevers were then subjected to piezo actuation [77]. The failure criterion for the thin film samples needed to be chosen differently since the thin film is attached to the substrate. Indeed, film fatigue damage is not catastrophic because the undamaged substrate maintains some mechanical integrity of the sample. Therefore, a crack in the Cu thin film does not lead to a failure of the whole sample. During fatigue, extrusions form and increase the roughness of the thin film surface, which can be detected by the surface reflectivity. The surface reflectivity was thus chosen as the fatigue failure criterion. The thin film lifetime data does not show a lot of scatter and follows a constant decrease in lifetime.

This can be explained by the large number of grains involved due to the small grain size of less than 1 µm in relation to the large sample size (1 mm wide cantilever) and the statistical grain size distribution.

Finally, the method described in this work focuses on the crack initiation and short crack growth in the early stages, and cannot be used to observe the evolution of long cracks due to the sample size and applied stress gradients. While the stress gradient at the sample surface across the 500 µm width is smaller than 2%, the stress gradient through the thickness is much higher and therefore can interfere with cracks longer than 20-30 µm. At this length, the cracks can reach far enough into the sample depth that the stress at the crack tip is reduced. This is not true for cracks that reach through the full width of the sample. In this case, the neutral axis is shifted as well, and the stress intensity at the crack tip does not change. Nevertheless, relative to the sample size, a 20-30 µm long crack is a mechanically rather severe crack, and the sample lifetime from this point on is rather short. In conclusion, even though the stress gradient in our micro-samples could have an impact on crack evolution in the later stages, it is not the case for the lifetime during crack initiation, which is a large part of the lifetime in the VHCF regime. However, in the lower HCF regime, the crack growth plays a considerable role and therefore the experiments presented here might underestimate the materials lifetime.

7 Summary

This thesis, as part of the DFG priority program SPP1466, had the ambition to investigate the fatigue mechanisms involved in the VHCF regime above 10^8 cycles. It specially focused on the fatigue mechanisms of small-scale Al and Cu by means of novel uniaxial and multiaxial custom-built resonant fatigue experimentation.

The development and implementation of an innovative experimental method has been completely fulfilled. The method allows for the investigation of the damage accumulation mechanisms, the precursors of the crack initiation in the case of small-scale fcc metals in the VHCF regime. The development and implementation of a novel custom-built resonant fatigue setup showed that the resonant frequency of bending micro-samples changes with increasing cycle number due to the accumulating fatigue damage. In the first regime up to $\Delta f/f_{initial} = -1x10^{-4}$, extrusions developed in many grains and consecutively micro-cracks initiated in single grains. Then, the crack initiation took place after a relative frequency decrease of about $\Delta f/f_{initial} = -1x10^{-3}$; where micro-cracks propagated through grain boundaries to form longer cracks. Finally, after a relative frequency decrease beyond roughly $\Delta f/f_{initial} = -5x10^{-3}$ long cracks propagated through the sample changing the resonant frequency at a fast rate.

Additional information on early damage formation has also been obtained. Mechanisms prior to crack initiation, such as plasticity induced surface roughening and slip band formation at a state where it appears in only a few grains, were observed. The accurate tracking of the fatigue damage evolution of pure nickel and pure copper is possible. The amount of damage has been optically estimated by calculating the amount of dark pixels on the sample surface, and has then been quantitatively correlated to the relative frequency decrease $\Delta f/f_{initial}$. A direct observation of cyclic hardening in grains can be tracked. When a low stress amplitude was applied, called 'subcritical amplitude' in this work, an increase of the resonant frequency has been observed.

The damage initiation investigation made with SEM and laser scanning microscopy showed that in many cases typical damage mechanism (roughness, slip bands, pores or micro-cracks) occurred in a few grains, but short cracks were not initiated even in the VHCF regime. The experimental results indicate that the effect can be explained by cyclic hardening, where the mobile dislocation segments are pinned. If the cyclic stress amplitude is too low, the strength of the pinning points cannot be overcome. Vacancy formation and oxidation formation have also been considered as possible explanations for this increase of the resonant frequency.

The implementation of a novel high-throughput method for the investigation of damage evolution from before the crack initiation to the short crack growth is work in progress. An in-situ optical acquisition system and an on-line image processing methodology are added to the new custom-built resonant fatigue setup. Individual grains and their neighbors on the whole sample surface will be observed and give even more insights on the early damage formation mechanism in the late HCF and VHCF regime.

The new experimental setup can also be used to experimentally define parameters needed for crack initiation models, such as the Tanaka-Mura or Chan. This will allow a more realistic prediction of lifetimes for applications structured at small-scales.

Finally, the crack initiation processes have been experimentally examined by this innovative method and supported by VHCF experiments of pure aluminum and pure copper. The fatigue behavior of samples and structures in the sub mm regime carries exciting new observations and opportunities for both basic science and applications.

8 Appendix

After test #	$\Delta f/f_{initial}$ [-]	Cycles [-]	After test #	$\Delta f/f_{initial}$ [-]	Cycles [-]
1	-1×10^{-4}	1.17×10^{5}	20	-4.5×10^{-3}	8.85×10^{7}
2	-2×10^{-4}	4.26×10^{5}	21	-4.9×10^{-3}	1.50×10^{8}
3	-3×10^{-4}	6.16×10^{5}	22	-5.3×10^{-3}	1.52×10^{8}
4	-4×10^{-4}	6.92×10^{5}	23	-5.7×10^{-3}	1.66×10^{8}
5	-5×10^{-4}	8.52×10^{5}	24	-6.1×10^{-3}	1.71×10^{8}
6	-7×10^{-4}	1.50×10^{6}	25	-6.5×10^{-3}	1.77×10^{8}
7	-9×10^{-4}	2.16×10^{6}	26	-6.9×10^{-3}	1.82×10^{8}
8	-1.1×10^{-3}	4.24×10^{6}	27	-7.7×10^{-3}	1.99×10^{8}
9	-1.3×10^{-3}	9.11×10^{6}	28	-8.5×10^{-3}	2.18×10^{8}
10	-1.5×10^{-3}	1.62×10^{7}	29	-9.3×10^{-3}	2.33×10^{8}
11	-1.7×10^{-3}	1.70×10^{7}	30	-1.01×10^{-2}	2.50×10^{8}
12	-1.9×10^{-3}	1.82×10^{7}	31	-1.11×10^{-3}	2.75×10^{8}
13	-2.1×10^{-3}	2.65×10^{7}	32	-1.21×10^{-3}	2.99×10^{8}
14	-2.3×10^{-3}	2.83×10^{7}	33	-1.31×10^{-3}	3.20×10^{8}
15	-2.5×10^{-3}	3.43×10^{7}	34	-1.41×10^{-3}	3.42×10^{8}
16	-2.9×10^{-3}	4.27×10^{7}	35	-1.6×10^{-3}	3.96×10^{8}
17	-3.3×10^{-3}	4.80×10^{7}	36	-1.8×10^{-3}	4.62×10^{8}
18	-3.7×10^{-3}	6.40×10^{7}	37	-2.2×10^{-3}	7.53×10^{8}
19	-4.1×10^{-3}	7.72×10^{7}			

Tab. 8.1: Cu44: Correlation between amount of damage and frequency change.

After test #	Relative pixel amount between the gray levels 0 and 50 [-]	Relative pixel amount between the gray levels 0 and 100 [-]	Relative pixel amount between the gray levels 0 and 150 [-]
1	0.000	0.000	0.000
2	0.041	0.036	0.011
3	0.029	0.009	-0.008
4	0.058	0.042	0.015
5	0.062	0.046	0.008
6	0.062	0.047	0.025
7	0.062	0.038	-0.001
8	0.058	0.030	0.005
9	0.012	0.019	-0.012
10	0.058	0.036	-0.004
11	0.058	0.034	-0.007
12	0.058	0.038	0.001
13	0.058	0.065	0.022
14	0.083	0.078	0.023
15	0.054	0.055	0.015
16	0.066	0.078	0.037
17	0.079	0.089	0.048
18	0.074	0.114	0.070
19	0.099	0.116	0.075
20	0.066	0.118	0.077
21	0.132	0.209	0.137
22	0.153	0.237	0.149
23	0.136	0.235	0.153
24	0.103	0.231	0.153
25	0.107	0.241	0.178
26	0.132	0.264	0.182
27	0.165	0.268	0.200
28	0.227	0.321	0.231
29	0.223	0.336	0.242
30	0.240	0.362	0.256
31	0.285	0.391	0.269
32	0.347	0.446	0.298
33	0.401	0.446	0.316
34	0.318	0.457	0.338
35	0.380	0.490	0.367
36	0.446	0.600	0.445
37	0.971	0.816	0.633

Tab. 8.2: Cu44 – Relative pixel amount between certain gray levels.

After test #	Relative pixel amount between the gray levels 0 and 50 [-]	Relative pixel amount between the gray levels 0 and 100 [-]	Relative pixel amount between the gray levels 0 and 150 [-]
1	0.000	0.000	0.000
4	0.000	0.000	0.006
7	0.000	0.002	0.029
10	0.003	0.118	0.434
20	0.011	0.571	0.776
30	0.113	0.914	0.945
37	1.000	0.999	0.979

Tab. 8.3: Cu44 – Relative pixel amount between certain gray levels.

Sample	Thickness [µm]	Stress [MPa]	Cycles [-]	$\Delta f/f_{initial}$ [-]
1	210	20.6	2.97301×10^5	-2×10^{-3}
12	174.65	11.94	4.40635×10^4	-2×10^{-3}
22	203.57	16.44	1.32×10^5	-2×10^{-3}
23	204.29	13.19	6.18×10^7	-2×10^{-3}
29	213.47	18.66	3.39×10^5	-2×10^{-3}
30	216.24	16.56	3.68×10^5	-2×10^{-3}
33	220.4	13.92	4.06448×10^6	-2×10^{-3}
40	221.78	12.52	3.85132×10^6	-2×10^{-3}
41	227.33	12.83	5.43×10^8	-2×10^{-3}
43	221.78	12.08	2.17×10^8	-2×10^{-3}
11	213.47	9.12	1.86469×10^7	2×10^{-3}
20	216.24	10.92	2.83×10^8	2×10^{-3}
24	210	8.72	3.86223×10^7	2×10^{-3}
34	221.78	10.9	2.36×10^7	2×10^{-3}
36	221.78	11.63	9.92318×10^7	2×10^{-3}
42	223.17	11.7	1.16×10^8	2×10^{-3}

Tab. 8.4: Al – micro-bending fatigue tests.

Sample	Thickness [µm]	Stress [MPa]	Cycles [-]	$\Delta f/f_{initial}$ [-]
10	210	15.35	4.48433×10^{6}	-2×10^{-3}
13	166.34	13.02	2.560345×10^{5}	-2×10^{-3}
19	210	10.56	6.80818×10^{7}	-2×10^{-3}
25	196.83	9.92	2.6222×10^{7}	-2×10^{-3}
26	203.76	12.08	1.67467×10^{6}	-2×10^{-3}
31	188.51	8.27	3.52×10^{8}	-2×10^{-3}
35	217.62	7	3.18028×10^{9}	-2×10^{-3}
14	210	3.46	1.08921×10^{8}	2×10^{-3}
38	220.4	5.29	8.924×10^{7}	2×10^{-3}

Tab. 8.5: Al – micro-torsion fatigue tests.

Sample	Thickness [µm]	Stress [MPa]	Cycles [-]	$\Delta f/f_{initial}$ [-]
5	185.74	165.46	7.69×10^{6}	-2×10^{-3}
8	181.58	155.02	7.97285×10^{4}	-2×10^{-3}
10	184.36	78.87	1.04614×10^{6}	-2×10^{-3}
13	180.16	148.2	3.41254×10^{6}	-2×10^{-3}
17	181.58	161.75	2.79298×10^{6}	-2×10^{-3}
26	181.58	114.23	1.203065×10^{5}	-2×10^{-3}
27	182.87	137.67	3.71463×10^{6}	-2×10^{-3}
32	184.36	101.23	2.18529×10^{6}	-2×10^{-3}
33	195.45	137.99	3.87159×10^{7}	-2×10^{-3}
34	170.5	64.36	7.2023×10^{9}	-2×10^{-3}
35	162.18	80.03	2.57205×10^{8}	-2×10^{-3}
36	163.56	80.34	4.00868×10^{7}	-2×10^{-3}
1	166.34	29.64	1.75×10^{7}	1.72×10^{-3}
14	182.97	39.16	1.00549×10^{7}	2×10^{-3}
15	180.6	57.69	3.14174×10^{7}	9×10^{-4}
16	171.88	79.44	3.22882×10^{7}	1.7×10^{-3}
31	185.74	25.78	1.58387×10^{10}	2×10^{-3}

Tab. 8.6: Cu – micro-bending fatigue tests.

Sample	Thickness [µm]	Stress [MPa]	Cycles [-]	$\Delta f/f_{initial}$ [-]
25	141.3	65.27	4.07×10^{6}	-7.4×10^{-4}
12	176.04	91.42	6.03×10^{7}	1.477×10^{-3}

Tab. 8.7: Cu – micro-torsion fatigue tests.

Sample	Thickness [µm]	Stress [MPa]	Cycles [-]	$\Delta f/f_{initial}$ [-]
2	180.16	73.06	2.15×10^{7}	-2×10^{-3}

Tab. 8.8: Cu – micro multiaxial fatigue test.

Cycles [-]	$\Delta f/f_{initial}$ [-]	Cycles [-]	$\Delta f/f_{initial}$ [-]
3.99×10^{8}	1.78×10^{-3}	4.22×10^{9}	1.61×10^{-4}
6.73×10^{8}	1.77×10^{-3}	4.36×10^{9}	1.44×10^{-4}
6.93×10^{8}	1.27×10^{-3}	4.94×10^{9}	-5.98×10^{-5}
1.02×10^{9}	1.33×10^{-3}	5.23×10^{9}	-1.21×10^{-4}
1.07×10^{9}	1.25×10^{-3}	5.31×10^{9}	-3.36×10^{-4}
1.36×10^{9}	1.33×10^{-3}	5.72×10^{9}	-4.61×10^{-4}
1.39×10^{9}	8.63×10^{-4}	5.76×10^{9}	-6.10×10^{-4}
1.58×10^{9}	7.77×10^{-4}	6.23×10^{9}	-7.81×10^{-4}
1.85×10^{9}	1.03×10^{-3}	6.59×10^{9}	-9.57×10^{-4}
2.19×10^{9}	9.03×10^{-4}	6.74×10^{9}	-1.18×10^{-3}
2.65×10^{9}	7.37×10^{-4}	6.90×10^{9}	-1.44×10^{-3}
3.07×10^{9}	4.70×10^{-4}	7.06×10^{9}	-1.71×10^{-3}
3.31×10^{9}	3.08×10^{-4}	7.19×10^{9}	-1.84×10^{-3}
4.01×10^{9}	3.22×10^{-4}	7.20×10^{9}	-2×10^{-3}

Tab. 8.9: Cu34 – The relative resonant frequency change is given for different number of cycles along the whole test. The frequency first increased before decreasing until reaching the criterion: $\Delta f/f_{initial} = -2 \times 10^{-3}$.

9 References

[1] W.A.J. Albert. Über Treibseile am Harz. *Archive für Mineralogie Geognosie Bergbau und Hüttenkunde*, 10:215–234, 1838.

[2] W. Rankine. On the mechanical action of heat, especially in gases and vapours. *Transactions of the Royal Society of Edinburgh*, 20:147–190, 1850.

[3] A. Wöhler. Theorie rechteckiger eiserner Brückenbalken mit Gitterwänden und mit Blechwänden. *Zeitschrift für Bauwesen*, 5:121–166, 1855.

[4] L.F.J. Coffin. A study of the effect of cyclic thermal stresses on a ductile metal. 1953.

[5] S.S. Manson. Behavior of materials under conditions of thermal stress. *National advisory commission on aeronautics*, 1170, 1954.

[6] S. Suresh. *Fatigue of materials*. Cambridge university press, 1998.

[7] H. Mughrabi. Fatigue, an everlasting materials problem - still en vogue. *Procedia Engineering*, 2(1):3–26, Apr 2010.

[8] H. Mughrabi. Cyclic slip irreversibilities and the evolution of fatigue damage. *Metallurgical and Materials Transactions B*, 40(4):431–453, Apr 2009.

[9] H. Mughrabi. On 'multi-stage' fatigue life diagrams and the relevant life-controlling mechanisms in ultrahigh-cycle fatigue. *Fatigue & Fracture of Engineering Materials and Structures*, 25(8-9):755–764, Sep 2002.

[10] C. Stöcker, M. Zimmermann, and H.-J. Christ. Localized cyclic deformation and corresponding dislocation arrangements of polycrystalline Ni-base superalloys and pure nickel in the VHCF regime. *International Journal of Fatigue*, 33(1):2–9, Jan 2011.

[11] A.G. Guy. *Introduction to Materials Science (McGraw-Hill series in materials science and engineering)*. McGraw-Hill Inc.,US, 1972.

[12] B.-T. Ma and C. Laird. Overview of fatigue behavior in copper single crystals - II. Population, size distribution and growth kinetics of Stage I cracks for tests at constant strain amplitude. *Acta Metallurgica*, 37(2):337–348, Feb 1989.

[13] J.C. Figueroa and C. Laird. Crack initiation mechanisms in copper polycrystals cycled under constant strain amplitudes and in step tests. *Materials Science and Engineering*, 60(1):45–58, Jul 1983.

[14] W.H. Kim and C. Laird. Crack nucleation and stage I propagation in high strain fatigue - I. Microscopic and interferometric observations. *Acta Metallurgica*, 26(5):777–787, May 1978.

[15] V.A Lubarda, M.S Schneider, D.H Kalantar, B.A Remington, and M.A Meyers. Void growth by dislocation emission. *Acta Materialia*, 52(6):1397–1408, Apr 2004.

[16] Y. Yang, B.I. Imasogie, S.M. Allameh, B. Boyce, K. Lian, J. Lou, and W.O. Soboyejo. Mechanisms of fatigue in LIGA Ni MEMS thin films. *Materials Science and Engineering: A*, 444(1-2):39–50, Jan 2007.

[17] N. Thompson, N. Wadsworth, and N. Louat. Xi. The origin of fatigue fracture in copper. *Philosophical Magazine*, 1(2):113–126, Feb 1956.

[18] B.L. Boyce, J.R. Michael, and P.G. Kotula. Fatigue of metallic microdevices and the role of fatigue-induced surface oxides. *Acta Materialia*, 52(6):1609–1619, Apr 2004.

[19] E.K. Baumert, F. Sadeghi-Tohidi, E. Hosseinian, and O.N. Pierron. Fatigue-induced thick oxide formation and its role on fatigue crack initiation in Ni thin films at low temperatures. *Acta Materialia*, 67:156–167, Apr 2014.

[20] W.L. Shan, Y. Yang, K.T. Hillie, W.A. Jordaan, and W.O. Soboyejo. Role of oxide thickening in fatigue crack initiation in LIGA nickel (MEMS) thin films. *Materials Science and Engineering: A*, 561:434–440, Jan 2013.

[21] K.S. Chan. A microstructure-based fatigue-crack-initiation model. *Metallurgical and Materials Transactions A*, 34(1):43–58, Jan 2003.

[22] B.A. Cowles. Life prediction in high temperature environments: Overview of a current gas turbine engine approach. *Materials Science and Engineering: A*, 103(1):63–69, Aug 1988.

[23] M.R. Mitchell. Fundamentals of modern fatigue analysis for design. *Fatigue and Microstructure - American Society for Metals*, pages 385–437, 1979.

[24] A.S. Cheng and C. Laird. Fatigue life behavior of copper single crystals. Part II: model for crack nucleation in persistent slip bands. *Fatigue & Fracture of Engineering Materials & Structures*, 4:343–353, 1981.

[25] K. Tanaka and T. Mura. A dislocation model for fatigue crack initiation. *J. Appl. Mech.*, 48(1):97–103, 1981.

[26] K.. Tanaka and T. Mura. A theory of fatigue crack initiation at inclusions. *MTA*, 13(1):117–123, Jan 1982.

[27] S.E. Harvey, P.G. Marsh, and W.W. Gerberich. Atomic force microscopy and modeling of fatigue crack initiation in metals. *Acta Metallurgica et Materialia*, 42(10):3493–3502, Oct 1994.

[28] T. Mura and Y. Nakasone. A theory of fatigue crack initiation in solids. *J. Appl. Mech.*, 57(1):1–6, 1990.

[29] U. Krupp, A. Giertler, M. Söker, H. Fu, B. Dönges, H.-J. Christ, A. Hüsecken, U. Pietsch, C.-P. Fritzen, and W. Ludwig. The behavior of short fatigue cracks during Very High Cycle (VHCF) Fatigue of duplex stainless steel. *Engineering Fracture Mechanics*, Mar 2015.

[30] C. Müller-Bollenhagen, M. Zimmermann, and H.-J. Christ. Very high cycle fatigue behaviour of austenitic stainless steel and the effect of strain-induced martensite. *International Journal of Fatigue*, 32(6):936–942, Jun 2010.

[31] S Stanzl-Tschegg. Fatigue crack growth and thresholds at ultrasonic frequencies. *International Journal of Fatigue*, 28(11):1456–1464, Nov 2006.

[32] C. Bathias, L. Drouillac, and P. Le Francois. How and why the fatigue S-N curve does not approach a horizontal asymptote. *International Journal of Fatigue*, 23:143–151, 2001.

[33] K. Shiozawa, Y. Morii, S. Nishino, and L. Lu. Subsurface crack initiation and propagation mechanism in high-strength steel in a very high cycle fatigue regime. *International Journal of Fatigue*, 28(11):1521–1532, Nov 2006.

[34] H.-J. Christ and H. Mughrabi. Cyclic stress-strain response ande microstructure under variable amplitude loading. *Fatigue & Fracture of Engineering Materials and Structures*, 19(2-3):335–348, Feb 1996.

[35] C. Eberl, R. Spolenak, O. Kraft, F. Kubat, W. Ruile, and E. Arzt. Damage analysis in Al thin films fatigued at ultrahigh frequencies. *Journal of Applied Physics*, 99(11):113501, 2006.

[36] C. Eberl, D. Courty, A. Walcker, and O. Kraft. Stress-gradient induced fatigue at ultra high frequencies in sub micron thin metal films. *International Journal of Materials Research*, 103(1):80–86, Jan 2012.

[37] R. Wang, H. Mughrabi, S. McGovern, and M. Rapp. Fatigue of copper single crystals in vacuum and in air. I: Persistent slip bands and dislocation microstructures. *Materials Science and Engineering*, 65(2):219–233, Aug 1984.

[38] A. Weidner, D. Amberger, F. Pyczak, B. Schönbauer, S. Stanzl-Tschegg, and H. Mughrabi. Fatigue damage in copper polycrystals subjected to ultrahigh-cycle fatigue below the PSB threshold. *International Journal of Fatigue*, 32(6):872–878, Jun 2010.

[39] S Stanzl-Tschegg, H. Mughrabi, and B. Schönbauer. Life time and cyclic slip of copper in the VHCF regime. *International Journal of Fatigue*, 29(9-11):2050–2059, Sep 2007.

[40] N.L. Phung, V. Favier, N. Ranc, F. Valès, and H. Mughrabi. Very high cycle fatigue of copper: Evolution, morphology and locations of surface slip markings. *International Journal of Fatigue*, 63:68–77, Jun 2014.

[41] H. Mughrabi. Specific features and mechanisms of fatigue in the ultrahigh-cycle regime. *International Journal of Fatigue*, 28(11):1501–1508, Nov 2006.

[42] U. Krupp. *Fatigue Crack Propagation in Metals and Alloys: Microstructural Aspects and Modelling Concepts*. Wiley, 2007.

[43] S. Stanzl-Tschegg. Fracture mechanisms and fracture mechanics at ultrasonic frequencies. *Fatigue & Fracture of Engineering Materials & Structures*, 22(7):567–579, Jul 1999.

[44] I. Marines-Garcia, J. Doucet, and C. Bathias. Development of a new device to perform torsional ultrasonic fatigue testing. *International Journal of Fatigue*, 29(9-11):2094–2101, Sep 2007.

[45] Y. Furuya. Specimen size effects on gigacycle fatigue properties of high-strength steel under ultrasonic fatigue testing. *Scripta Materialia*, 58(11):1014–1017, Jun 2008.

[46] G.P. Zhang, C.A. Volkert, R. Schwaiger, P. Wellner, E. Arzt, and O. Kraft. Length-scale-controlled fatigue mechanisms in thin copper films. *Acta Materialia*, 54(11):3127–3139, Jun 2006.

[47] T. Straub, E.K. Baumert, C. Eberl, and O.N. Pierron. A method for probing the effects of conformal nanoscale coatings on fatigue crack initiation in electroplated Ni films. *Thin Solid Films*, 526:176–182, Dec 2012.

[48] J.I. Latham, W.R. Shreve, N.J. Tolar, and P.B. Ghate. Improved metallization for surface acoustic wave devices. *Thin Solid Films*, 64(1):9–15, Nov 1979.

[49] R. Schwaiger and O. Kraft. Size effects in the fatigue behavior of thin Ag films. *Acta Materialia*, 51(1):195–206, Jan 2003.

[50] N. Walker and C.J. Beevers. A fatigue crack closure mechanism in Titanium. *Fatigue & Fracture of Engineering Materials and Structures*, 1(1):135–148, Jan 1979.

[51] J. Charrier and R. Roux. Evolution of damage fatigue by electrical measure on smooth cylindrical specimens. *Nondestructive Testing and Evaluation*, 6(2):113–124, Jan 1991.

[52] B. Sun, L. Yang, and Y. Guo. A high-cycle fatigue accumulation model based on electrical resistance for structural steels. *Fatigue & Fracture of Engineering Materials and Structures*, 30(11):1052–1062, Nov 2007.

[53] D. Wagner, N. Ranc, C. Bathias, and P.C. Paris. Fatigue crack initiation detection by an infrared thermography method. *Fatigue & Fracture of Engineering Materials & Structures*, 33(1):12–21, Nov 2009.

[54] P. Starke, F. Walther, and D. Eifler. PHYBAL - A new method for lifetime prediction based on strain, temperature and electrical measurements. *International Journal of Fatigue*, 28(9):1028–1036, Sep 2006.

[55] A. Kumar, C.J. Torbet, T.M. Pollock, and Jones J.W. In situ characterization of fatigue damage evolution in a cast Al alloy via nonlinear ultrasonic measurements. *Acta Materialia*, 58(6):2143–2154, Apr 2010.

[56] M. Funk and C. Eberl. Investigations on the fatigue behavior of nanocrystalline metals. *Key Engineering Materials*, 465:207–210, Jan 2011.

[57] C. Mattheck. Teacher tree: The evolution of notch shape optimization from complex to simple. *Engineering Fracture Mechanics*, 73(12):1732–1742, Aug 2006.

[58] K.J. Aström and T. Hägglund. *Advanced PID Control*. ISA - The Instrumentation, Systems, and Automation Society; Research Triangle Park, NC 27709, 2006.

[59] T.L. Anderson and T.L. Anderson. *Fracture Mechanics: Fundamentals and Applications, Third Edition*. Taylor & Francis, 2005.

[60] M. Zimmermann. Diversity of damage evolution during cyclic loading at very high numbers of cycles. *International Materials Reviews*, 57(2):73–91, Mar 2012.

[61] M. Koster, G. Wagner, and D. Eifler. Cyclic deformation behavior of a medium carbon steel in the VHCF regime. *Procedia Engineering*, 2(1):2189–2197, Apr 2010.

[62] U. Rabe, T. Helfen, M. Weikert, S. Hirsekorn, H.G. Herrmann, C. Boller, D. Backe, F. Balle, and D. Eifler. Nonlinear ultrasonic testing of carbon fibre reinforced plastics in the very high cycle fatigue regime. *Acoustical Society of America - Proceedings of Meetings on Acoustics*, 16(1):045004, 2012.

[63] Y. Yang, S. Allameh, J. Lou, B. Imasogie, B.L. Boyce, and W.O. Soboyejo. Fatigue of LIGA Ni micro-electro-mechanical system thin films. *Metallurgical and Materials Transactions A*, 38(13):2340–2348, Aug 2007.

[64] Y. Yang, S. Allameh, B. Boyce, K.S. Chan, and W.O. Soboyejo. An experimental study of fracture of LIGA Ni micro-electro-mechanical systems thin films. *Metallurgical and Materials Transactions A*, 38(6):1223–1230, Jun 2007.

[65] S. Burger. *High Cycle Fatigue of Al and Cu Thin Films by a Novel High-Throughput Method*. Schriftenreihe des Instituts für Angewandte Materialien, Karlsruher Institut für Technologie. KIT Scientific Publishing, 2013.

[66] E.A. Mitscherlich. *Das Gesetz des Minimums und das Gesetz des abnehmenden Bodenertrags*, volume 38: p. 537-552. Landwirtschaftliche Jahrbücher, 1909.

[67] D. Weygand, L.H. Friedman, E.vd Giessen, and A. Needleman. Aspects of boundary-value problem solutions with three-dimensional dislocation dynamics. *Modelling Simul. Mater. Sci. Eng.*, 10(4):437–468, Jun 2002.

[68] F. Siska, S. Forest, P. Gumbsch, and D. Weygand. Finite element simulations of the cyclic elastoplastic behaviour of copper thin films. *Modelling Simul. Mater. Sci. Eng.*, 15(1):S217–S238, Dec 2006.

[69] F. Siska, D. Weygand, S. Forest, and P. Gumbsch. Comparison of mechanical behaviour of thin film simulated by discrete dislocation dynamics and continuum crystal plasticity. *Computational Materials Science*, 45(3):793–799, May 2009.

[70] D. Kiener, C. Motz, W. Grosinger, D. Weygand, and R. Pippan. Cyclic response of copper single crystal micro-beams. *Scripta Materialia*, 63(5):500–503, Sep 2010.

[71] H.W. Hoeppel, L. May, M. Prell, and M. Goeken. Influence of grain size and precipitation state on the fatigue lives and deformation mechanisms of CP aluminium and AA6082 in the VHCF-regime. *International Journal of Fatigue*, 33(1):10–18, Jan 2011.

[72] S. Stanzl-Tschegg and B. Schönbauer. Mechanisms of strain localization, crack initiation and fracture of polycrystalline copper in the VHCF regime. *International Journal of Fatigue*, 32(6):886–893, Jun 2010.

[73] A.W. Thompson and W.A. Backofen. The effect of grain size on fatigue. *Acta Metallurgica*, 19(7):597–606, Jul 1971.

[74] L.F. Mondolfo. *Aluminum Alloys: Structure and Properties*. Elsevier Science, 2013.

[75] T. Kennerknecht. *Fatigue of Micro Molded Materials - Aluminum Bronze and Yttria Stabilized Zirconia*. Schriftenreihe des Instituts fuer Angewandte Materialien, Karlsruher Institut fuer Technologie. KIT Scientific Publishing, 2014.

[76] G. Baumeister, B. Okolo, and J. Rögner. Microcasting of al bronze: influence of casting parameters on the microstructure and the mechanical properties. *Microsystem Technologies*, 14(9-11):1647–1655, May 2008.

[77] S. Burger, C. Eberl, A. Siegel, A. Ludwig, and O. Kraft. A novel high-throughput fatigue testing method for metallic thin films. *Sci. Technol. Adv. Mater.*, 12(5):054202, Sep 2011.

Schriftenreihe
des Instituts für Angewandte Materialien

ISSN 2192-9963

Die Bände sind unter www.ksp.kit.edu als PDF frei verfügbar
oder als Druckausgabe bestellbar.

Band 1 Prachai Norajitra
Divertor Development for a Future Fusion Power Plant. 2011
ISBN 978-3-86644-738-7

Band 2 Jürgen Prokop
Entwicklung von Spritzgießsonderverfahren zur Herstellung von Mikrobauteilen durch galvanische Replikation. 2011
ISBN 978-3-86644-755-4

Band 3 Theo Fett
New contributions to R-curves and bridging stresses – Applications of weight functions. 2012
ISBN 978-3-86644-836-0

Band 4 Jérôme Acker
Einfluss des Alkali/Niob-Verhältnisses und der Kupferdotierung auf das Sinterverhalten, die Strukturbildung und die Mikrostruktur von bleifreier Piezokeramik $(K_{0,5}Na_{0,5})NbO_3$. 2012
ISBN 978-3-86644-867-4

Band 5 Holger Schwaab
Nichtlineare Modellierung von Ferroelektrika unter Berücksichtigung der elektrischen Leitfähigkeit. 2012
ISBN 978-3-86644-869-8

Band 6 Christian Dethloff
Modeling of Helium Bubble Nucleation and Growth in Neutron Irradiated RAFM Steels. 2012
ISBN 978-3-86644-901-5

Band 7 Jens Reiser
Duktilisierung von Wolfram. Synthese, Analyse und Charakterisierung von Wolframlaminaten aus Wolframfolie. 2012
ISBN 978-3-86644-902-2

Band 8 Andreas Sedlmayr
Experimental Investigations of Deformation Pathways in Nanowires. 2012
ISBN 978-3-86644-905-3

Band 9 Matthias Friedrich Funk
Microstructural stability of nanostructured fcc metals during cyclic deformation and fatigue. 2012
ISBN 978-3-86644-918-3

Band 10 Maximilian Schwenk
Entwicklung und Validierung eines numerischen Simulationsmodells zur Beschreibung der induktiven Ein- und Zweifrequenzrandschichthärtung am Beispiel von vergütetem 42CrMo4. 2012
ISBN 978-3-86644-929-9

Band 11 Matthias Merzkirch
Verformungs- und Schädigungsverhalten der verbundstranggepressten, federstahldrahtverstärkten Aluminiumlegierung EN AW-6082. 2012
ISBN 978-3-86644-933-6

Band 12 Thilo Hammers
Wärmebehandlung und Recken von verbundstranggepressten Luftfahrtprofilen. 2013
ISBN 978-3-86644-947-3

Band 13 Jochen Lohmiller
Investigation of deformation mechanisms in nanocrystalline metals and alloys by in situ synchrotron X-ray diffraction. 2013
ISBN 978-3-86644-962-6

Band 14 Simone Schreijäg
Microstructure and Mechanical Behavior of Deep Drawing DC04 Steel at Different Length Scales. 2013
ISBN 978-3-86644-967-1

Band 15 Zhiming Chen
Modelling the plastic deformation of iron. 2013
ISBN 978-3-86644-968-8

Band 16 Abdullah Fatih Çetinel
Oberflächendefektausheilung und Festigkeitssteigerung von niederdruckspritzgegossenen Mikrobiegebalken aus Zirkoniumdioxid. 2013
ISBN 978-3-86644-976-3

Band 17 Thomas Weber
Entwicklung und Optimierung von gradierten Wolfram/ EUROFER97-Verbindungen für Divertorkomponenten. 2013
ISBN 978-3-86644-993-0

Band 18 Melanie Senn
Optimale Prozessführung mit merkmalsbasierter Zustandsverfolgung. 2013
ISBN 978-3-7315-0004-9

Band 19 Christian Mennerich
Phase-field modeling of multi-domain evolution in ferromagnetic shape memory alloys and of polycrystalline thin film growth. 2013
ISBN 978-3-7315-0009-4

Band 20 Spyridon Korres
On-Line Topographic Measurements of Lubricated Metallic Sliding Surfaces. 2013
ISBN 978-3-7315-0017-9

Band 21 Abhik Narayan Choudhury
Quantitative phase-field model for phase transformations in multi-component alloys. 2013
ISBN 978-3-7315-0020-9

Band 22 Oliver Ulrich
Isothermes und thermisch-mechanisches Ermüdungsverhalten von Verbundwerkstoffen mit Durchdringungsgefüge (Preform-MMCs). 2013
ISBN 978-3-7315-0024-7

Band 23 Sofie Burger
High Cycle Fatigue of Al and Cu Thin Films by a Novel High-Throughput Method. 2013
ISBN 978-3-7315-0025-4

Band 24 Michael Teutsch
Entwicklung von elektrochemisch abgeschiedenem LIGA-Ni-Al für Hochtemperatur-MEMS-Anwendungen. 2013
ISBN 978-3-7315-0026-1

Band 25 Wolfgang Rheinheimer
Zur Grenzflächenanisotropie von $SrTiO_3$. 2013
ISBN 978-3-7315-0027-8

Band 26 Ying Chen
Deformation Behavior of Thin Metallic Wires under Tensile and Torsional Loadings. 2013
ISBN 978-3-7315-0049-0

Band 27 Sascha Haller
Gestaltfindung: Untersuchungen zur Kraftkegelmethode. 2013
ISBN 978-3-7315-0050-6

Band 28 Stefan Dietrich
Mechanisches Verhalten von GFK-PUR-Sandwichstrukturen unter quasistatischer und dynamischer Beanspruchung. 2013
ISBN 978-3-7315-0074-2

Band 29 Gunnar Picht
Einfluss der Korngröße auf ferroelektrische Eigenschaften dotierter $Pb(Zr_{1-x}Ti_x)O_3$ Materialien. 2013
ISBN 978-3-7315-0106-0

Band 30 Esther Held
Eigenspannungsanalyse an Schichtverbunden mittels inkrementeller Bohrlochmethode. 2013
ISBN 978-3-7315-0127-5

Band 31 Pei He
On the structure-property correlation and the evolution of Nanofeatures in 12-13.5% Cr oxide dispersion strengthened ferritic steels. 2014
ISBN 978-3-7315-0141-1

Band 32 Jan Hoffmann
Ferritische ODS-Stähle – Herstellung, Umformung und Strukturanalyse. 2014
ISBN 978-3-7315-0157-2

Band 33 Wiebke Sittel
Entwicklung und Optimierung des Diffusionsschweißens von ODS Legierungen. 2014
ISBN 978-3-7315-0182-4

Band 34 Osama Khalil
Isothermes Kurzzeitermüdungsverhalten der hoch-warmfesten Aluminium-Knetlegierung 2618A (AlCu2Mg1,5Ni). 2014
ISBN 978-3-7315-0208-1

Band 35 Magalie Huttin
Phase-field modeling of the influence of mechanical stresses on charging and discharging processes in lithium ion batteries. 2014
ISBN 978-3-7315-0213-5

Band 36 Christoph Hage
Grundlegende Aspekte des 2K-Metallpulverspritzgießens. 2014
ISBN 978-3-7315-0217-3

Band 37 Bartłomiej Albiński
Instrumentierte Eindringprüfung bei Hochtemperatur für die Charakterisierung bestrahlter Materialien. 2014
ISBN 978-3-7315-0221-0

Band 38 Tim Feser
Untersuchungen zum Einlaufverhalten binärer alpha-Messinglegierungen unter Ölschmierung in Abhängigkeit des Zinkgehaltes. 2014
ISBN 978-3-7315-0224-1

Band 39 Jörg Ettrich
Fluid Flow and Heat Transfer in Cellular Solids. 2014
ISBN 978-3-7315-0241-8

Band 40 Melanie Syha
Microstructure evolution in strontium titanate Investigated by means of grain growth simulations and x-ray diffraction contrast tomography experiments. 2014
ISBN 978-3-7315-0242-5

Band 41 Thomas Haas
Mechanische Zuverlässigkeit von gedruckten und gasförmig abgeschiedenen Schichten auf flexiblem Substrat. 2014
ISBN 978-3-7315-0250-0

Band 42 Aron Kneer
Numerische Untersuchung des Wärmeübertragungsverhaltens in unterschiedlichen porösen Medien. 2014
ISBN 978-3-7315-0252-4

Band 43 Manuel Feuchter
Investigations on Joule heating applications by multiphysical continuum simulations in nanoscale systems. 2014
ISBN 978-3-7315-0261-6

Band 44 Alexander Vondrous
Grain growth behavior and efficient large scale simulations of recrystallization with the phase-field method. 2014
ISBN 978-3-7315-0280-7

Band 45 Tobias Kennerknecht
Fatigue of Micro Molded Materials – Aluminum Bronze and Yttria Stabilized Zirconia. 2014
ISBN 978-3-7315-0293-7

Band 46 Christopher Scherr
Elektrochemisches Verhalten von Lithium-Schwefel-Zellen mit unterschiedlicher Kathodenstruktur. 2015
ISBN 978-3-7315-0296-8

Band 47 Konstantin Frölich
Der Decal-Prozess zur Herstellung katalysatorbeschichteter Membranen für PEM-Brennstoffzellen. 2015
ISBN 978-3-7315-0334-7

Band 48 Benedikt Haspel
Werkstoffanalytische Betrachtung der Eigenschaften von mittels neuartiger RTM-Fertigungsprozesse hergestellten glasfaserverstärkten Polymerverbunden. 2015
ISBN 978-3-7315-0337-8

Band 49 Marco Berghoff
Skalenübergreifende Modellierung und Optimierung vom atomistischen kristallinen Phasenfeldmodell bis zur mesoskopischen Phasenfeldmethode. 2015
ISBN 978-3-7315-0416-0

Band 50 Michael Selzer
Mechanische und Strömungsmechanische Topologieoptimierung mit der Phasenfeldmethode. 2016
ISBN 978-3-7315-0431-3

Band 51 Michael Mahler
Entwicklung einer Auswertemethode für bruchmechanische Versuche an kleinen Proben auf der Basis eines Kohäsivzonenmodells. 2016
ISBN 978-3-7315-0441-2

Band 52 Christoph Bohnert
Numerische Untersuchung des Verformungs- und Bruchverhaltens von einkristallinem Wolfram auf mikroskopischer Ebene. 2016
ISBN 978-3-7315-0444-3

Band 53 Stefan Guth
Schädigung und Lebensdauer von Nickelbasislegierungen unter thermisch-mechanischer Ermüdungsbeanspruchung bei verschiedenen Phasenlagen. 2016
ISBN 978-3-7315-0445-0

Band 54 Markus Klinsmann
The Effects of Internal Stress and Lithium Transport on Fracture in Storage Materials in Lithium-Ion Batteries. 2016
ISBN 978-3-7315-0455-9

Band 55 Thomas Straub
Experimental Investigation of Crack Initiation in Face-Centered Cubic Materials in the High and Very High Cycle Fatigue Regime. 2016
ISBN 978-3-7315-0471-9